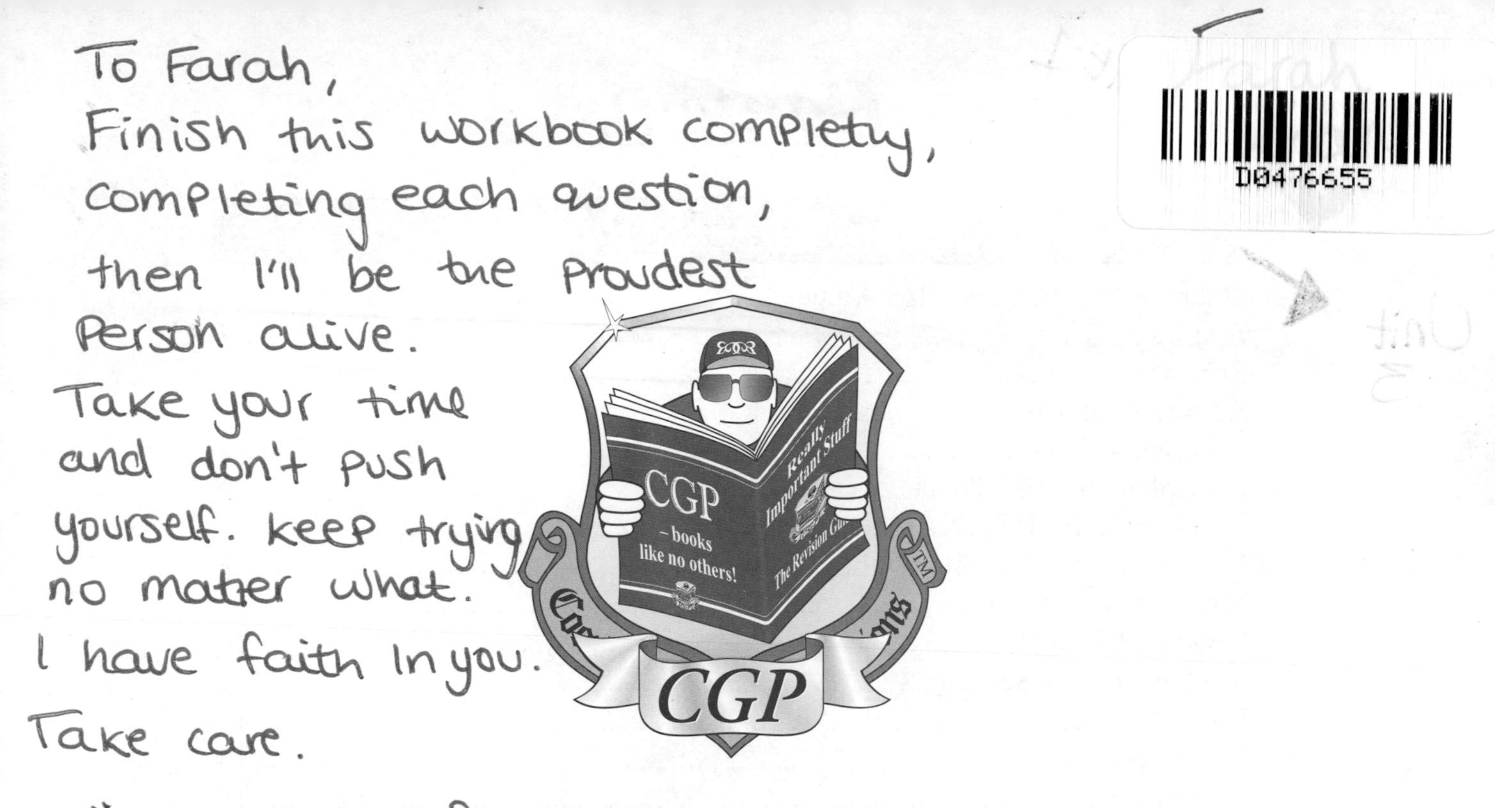

Nuff' Love -> Simran ♡x
A·K·A· Ducky :)

P.S. Have fun! :)

It's another Quality Book from CGP

This book is for anyone doing GCSE Mathematics at Foundation Level.

It contains lots of tricky questions designed to make you sweat — because that's the only way you'll get any better.

It's also got some daft bits in to try and make the whole experience at least vaguely entertaining for you.

What CGP is all about

Our sole aim here at CGP is to produce the highest quality books — carefully written, immaculately presented and dangerously close to being funny.

Then we work our socks off to get them out to you — at the cheapest possible prices.

Contents

Section One — Numbers

Section Two — Shapes and Area

Section Three — Measurements

Section Four — Angles and Geometry

Unit 3

Section Five — Handling Data

Unit 1

Section Six — Graphs

Unit 1
Unit 3

Section Seven — Algebra

Unit 2

Throughout the book, the more challenging questions are marked like this: **Q1**

* Unit 3 will have parts of everything.

Published by CGP
Illustrated by Ruso Bradley, Lex Ward and Ashley Tyson

From original material by Richard Parsons

Contributors:
Gill Allen, Margaret Carr, Barbara Coleman, JE Dodds, Mark Haslam, John Lyons, C McLoughlin, Gordon Rutter, John Waller, Janet West, Dave Williams, Philip Wood.

Updated by:
Sarah Blackwood, Rosie Gillham, Neil Hastings, Paul Jordin, Simon Little, Ali Palin, Julie Wakeling, Sarah Williams.

With thanks to Peter Caunter and Jane Towle for the proofreading.

ISBN: 978 1 84146 648 4

Groovy website: www.cgpbooks.co.uk
Printed by Elanders Ltd, Newcastle upon Tyne.
Jolly bits of clipart from CorelDRAW®

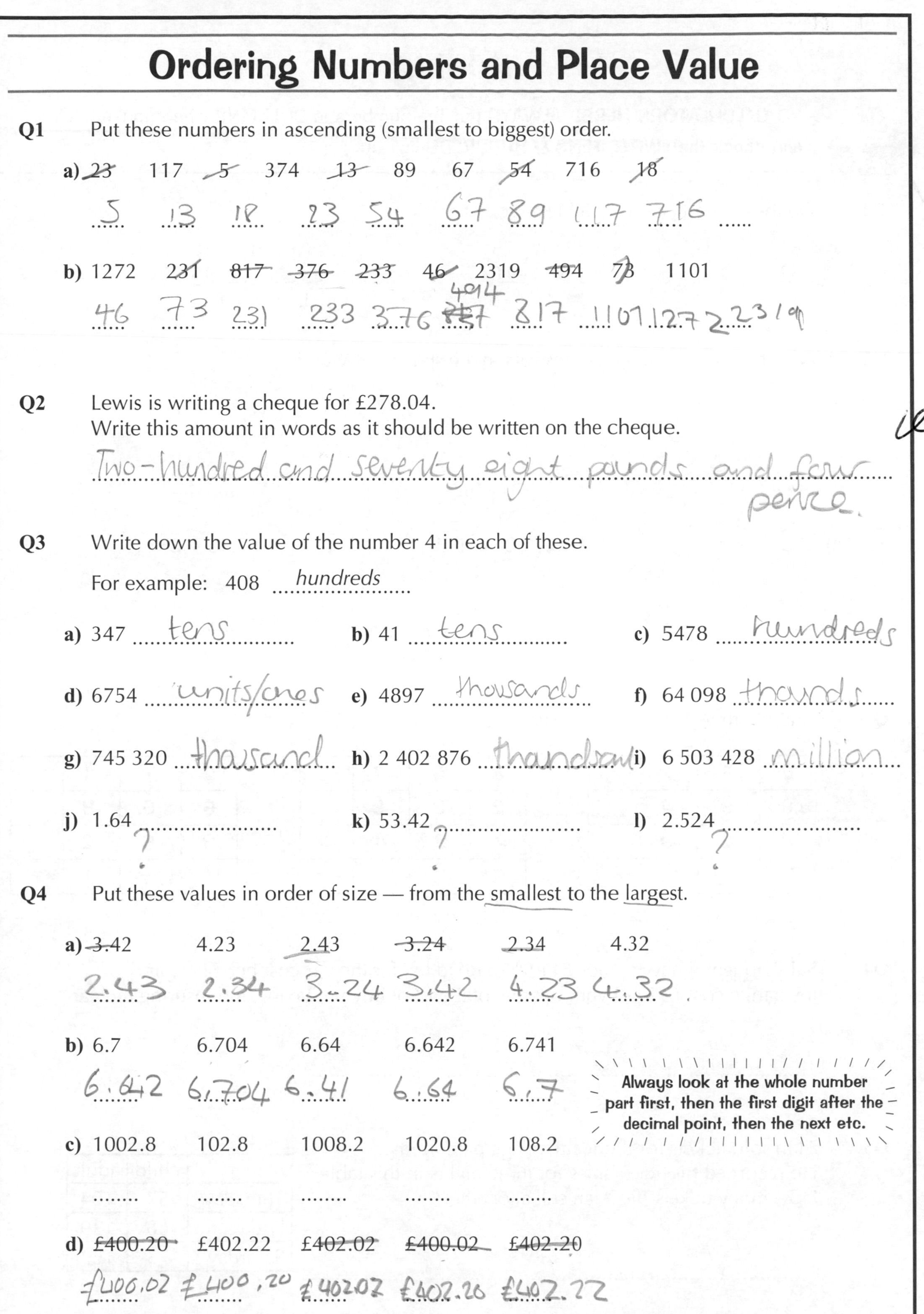

Ordering Numbers and Place Value

Q1 Put these numbers in ascending (smallest to biggest) order.

a) 23 117 5 374 13 89 67 54 716 18

....

b) 1272 231 817 376 233 46 2319 494 73 1101

....

Q2 Lewis is writing a cheque for £278.04.
Write this amount in words as it should be written on the cheque.

....

Q3 Write down the value of the number 4 in each of these.

For example: 408 *hundreds*

a) 347 .... **b)** 41 .... **c)** 5478 ....

d) 6754 .... **e)** 4897 .... **f)** 64 098 ....

g) 745 320 .... **h)** 2 402 876 .... **i)** 6 503 428 ....

j) 1.64 .... **k)** 53.42 .... **l)** 2.524 ....

Q4 Put these values in order of size — from the smallest to the largest.

a) 3.42 4.23 2.43 3.24 2.34 4.32

....

b) 6.7 6.704 6.64 6.642 6.741

....

Always look at the whole number part first, then the first digit after the decimal point, then the next etc.

c) 1002.8 102.8 1008.2 1020.8 108.2

....

d) £400.20 £402.22 £402.02 £400.02 £402.20

....

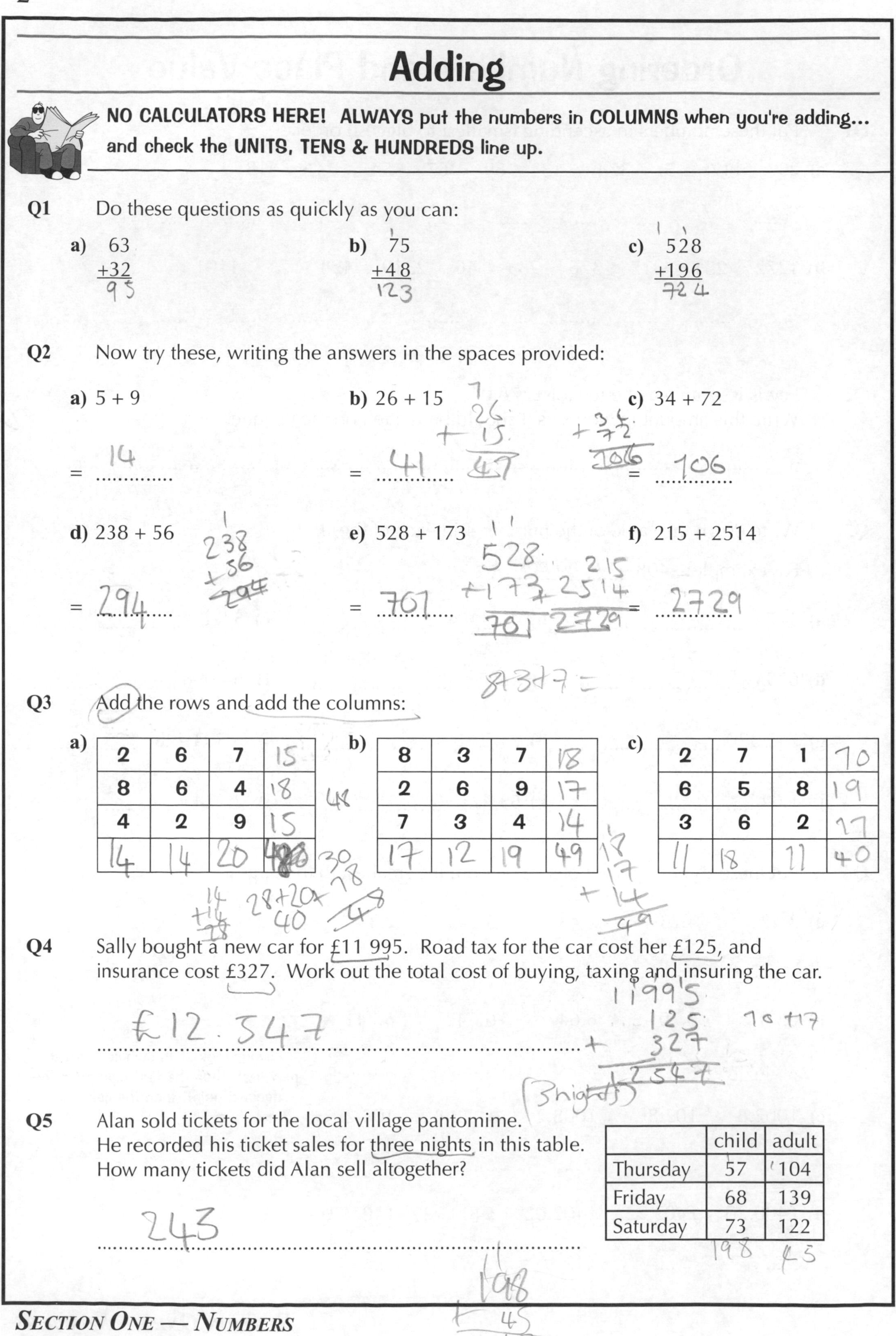

Adding

NO CALCULATORS HERE! ALWAYS put the numbers in COLUMNS when you're adding... and check the UNITS, TENS & HUNDREDS line up.

Q1 Do these questions as quickly as you can:

a) 63 + 32

b) 75 + 48

c) 528 + 196

Q2 Now try these, writing the answers in the spaces provided:

a) 5 + 9 = ...

b) 26 + 15 = ...

c) 34 + 72 = ...

d) 238 + 56 = ...

e) 528 + 173 = ...

f) 215 + 2514 = ...

Q3 Add the rows and add the columns:

a)

2	6	7	
8	6	4	
4	2	9	

b)

8	3	7	
2	6	9	
7	3	4	

c)

2	7	1	
6	5	8	
3	6	2	

Q4 Sally bought a new car for £11 995. Road tax for the car cost her £125, and insurance cost £327. Work out the total cost of buying, taxing and insuring the car.

....

Q5 Alan sold tickets for the local village pantomime. He recorded his ticket sales for three nights in this table. How many tickets did Alan sell altogether?

	child	adult
Thursday	57	104
Friday	68	139
Saturday	73	122

....

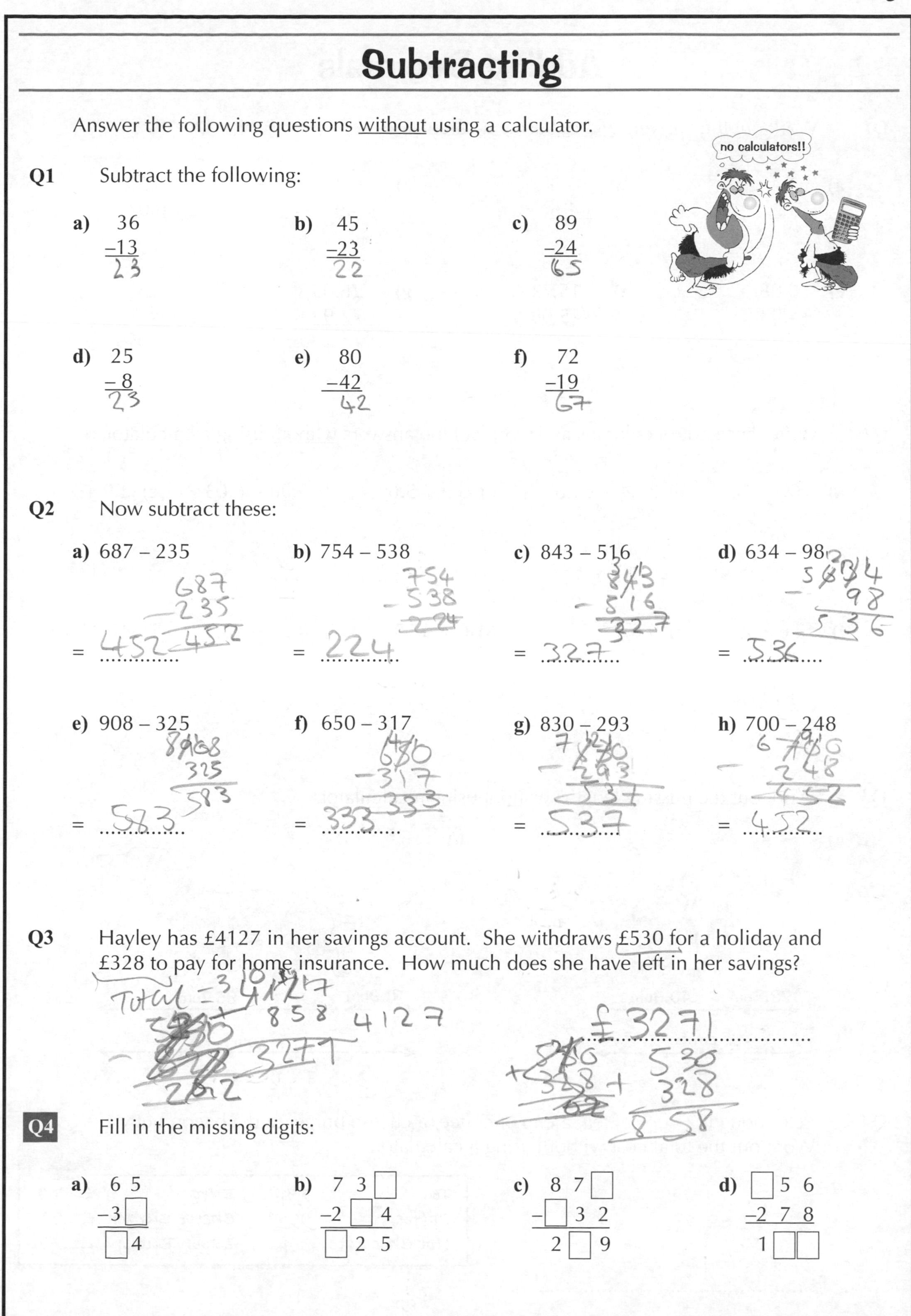

Subtracting

Answer the following questions <u>without</u> using a calculator.

Q1 Subtract the following:

a) 36 − 13

b) 45 − 23

c) 89 − 24

d) 25 − 8

e) 80 − 42

f) 72 − 19

Q2 Now subtract these:

a) 687 – 235 = ...

b) 754 – 538 = ...

c) 843 – 516 = ...

d) 634 – 98 = ...

e) 908 – 325 = ...

f) 650 – 317 = ...

g) 830 – 293 = ...

h) 700 – 248 = ...

Q3 Hayley has £4127 in her savings account. She withdraws £530 for a holiday and £328 to pay for home insurance. How much does she have left in her savings?

....

Q4 Fill in the missing digits:

a) 6 5 − 3☐ = ☐4

b) 7 3 ☐ − 2 ☐ 4 = ☐ 2 5

c) 8 7 ☐ − ☐ 3 2 = 2 ☐ 9

d) ☐ 5 6 − 2 7 8 = 1 ☐ ☐

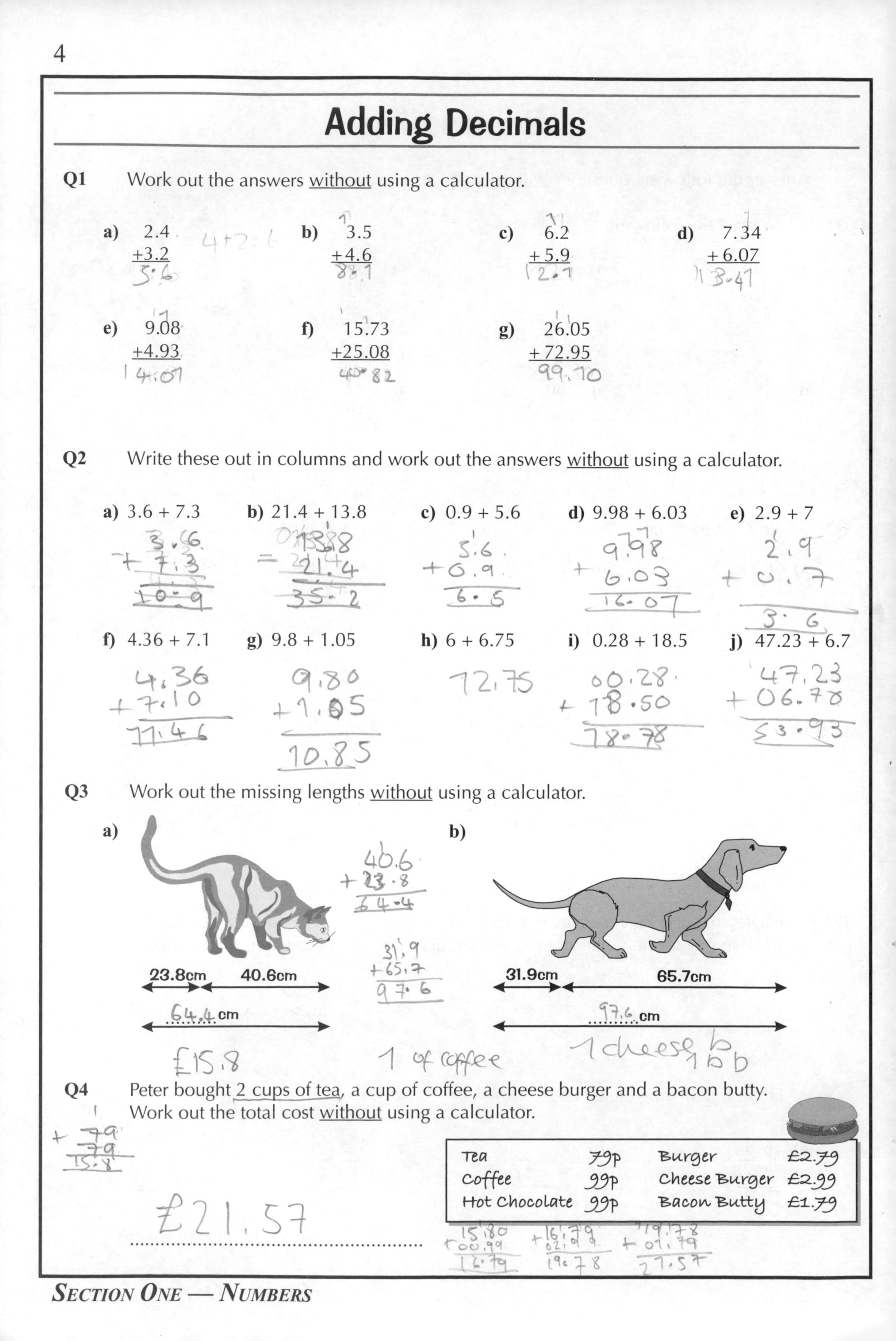

Adding Decimals

Q1 Work out the answers without using a calculator.

a) 2.4 +3.2 **b)** 3.5 +4.6 **c)** 6.2 +5.9 **d)** 7.34 +6.07

e) 9.08 +4.93 **f)** 15.73 +25.08 **g)** 26.05 +72.95

Q2 Write these out in columns and work out the answers without using a calculator.

a) 3.6 + 7.3 **b)** 21.4 + 13.8 **c)** 0.9 + 5.6 **d)** 9.98 + 6.03 **e)** 2.9 + 7

f) 4.36 + 7.1 **g)** 9.8 + 1.05 **h)** 6 + 6.75 **i)** 0.28 + 18.5 **j)** 47.23 + 6.7

Q3 Work out the missing lengths without using a calculator.

a)

b)

Q4 Peter bought 2 cups of tea, a cup of coffee, a cheese burger and a bacon butty.
Work out the total cost without using a calculator.

Tea	79p	Burger	£2.79
Coffee	99p	Cheese Burger	£2.99
Hot Chocolate	99p	Bacon Butty	£1.79

....

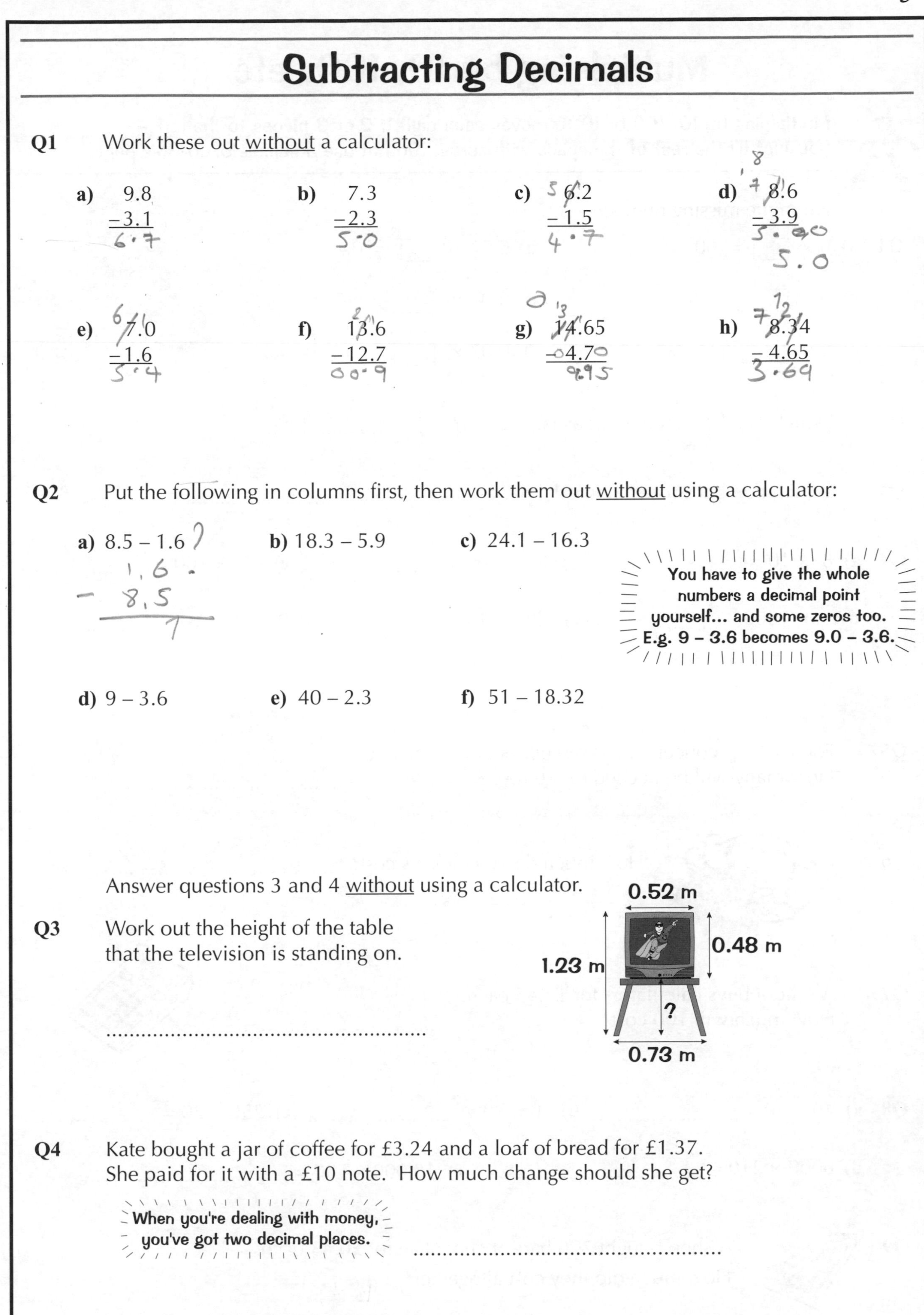

Subtracting Decimals

Q1 Work these out without a calculator:

a) 9.8 − 3.1 **b)** 7.3 − 2.3 **c)** 6.2 − 1.5 **d)** 8.6 − 3.9

e) 7.0 − 1.6 **f)** 13.6 − 12.7 **g)** 14.65 − 4.70 **h)** 8.34 − 4.65

Q2 Put the following in columns first, then work them out without using a calculator:

a) 8.5 – 1.6 **b)** 18.3 – 5.9 **c)** 24.1 – 16.3

d) 9 – 3.6 **e)** 40 – 2.3 **f)** 51 – 18.32

You have to give the whole numbers a decimal point yourself... and some zeros too. E.g. 9 – 3.6 becomes 9.0 – 3.6.

Answer questions 3 and 4 without using a calculator.

Q3 Work out the height of the table that the television is standing on.

....

Q4 Kate bought a jar of coffee for £3.24 and a loaf of bread for £1.37. She paid for it with a £10 note. How much change should she get?

When you're dealing with money, you've got two decimal places.

....

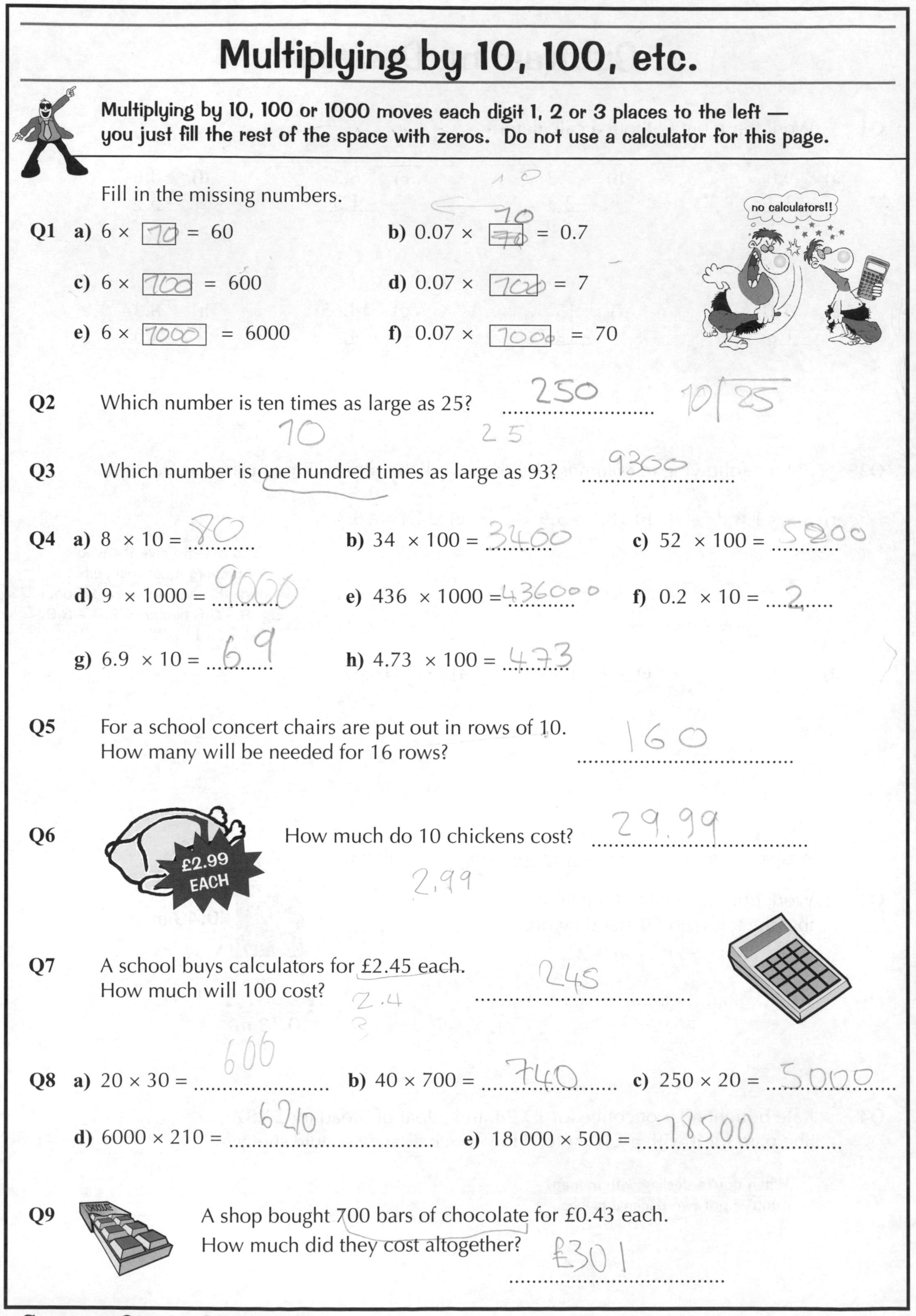

Multiplying by 10, 100, etc.

Multiplying by 10, 100 or 1000 moves each digit 1, 2 or 3 places to the left — you just fill the rest of the space with zeros. Do not use a calculator for this page.

Fill in the missing numbers.

Q1 **a)** 6 × ☐ = 60 **b)** 0.07 × ☐ = 0.7

c) 6 × ☐ = 600 **d)** 0.07 × ☐ = 7

e) 6 × ☐ = 6000 **f)** 0.07 × ☐ = 70

Q2 Which number is ten times as large as 25? ....

Q3 Which number is one hundred times as large as 93? ....

Q4 **a)** 8 × 10 = ... **b)** 34 × 100 = ... **c)** 52 × 100 = ...

d) 9 × 1000 = ... **e)** 436 × 1000 = ... **f)** 0.2 × 10 = ...

g) 6.9 × 10 = ... **h)** 4.73 × 100 = ...

Q5 For a school concert chairs are put out in rows of 10.
How many will be needed for 16 rows? ....

Q6 How much do 10 chickens cost? ....

Q7 A school buys calculators for £2.45 each.
How much will 100 cost? ....

Q8 **a)** 20 × 30 = ... **b)** 40 × 700 = ... **c)** 250 × 20 = ...

d) 6000 × 210 = ... **e)** 18 000 × 500 = ...

Q9 A shop bought 700 bars of chocolate for £0.43 each.
How much did they cost altogether? ....

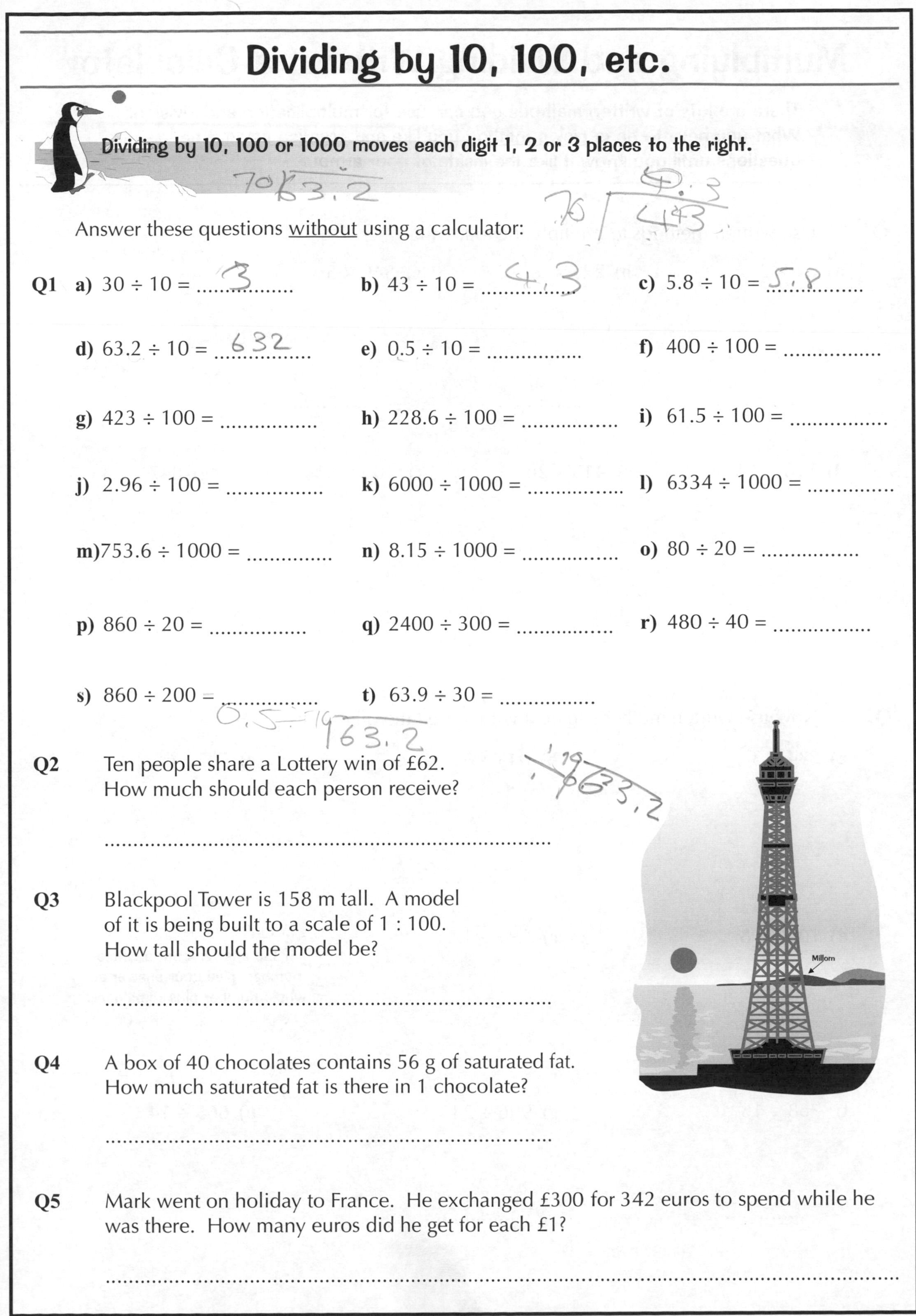

Dividing by 10, 100, etc.

Dividing by 10, 100 or 1000 moves each digit 1, 2 or 3 places to the right.

Answer these questions <u>without</u> using a calculator:

Q1 **a)** $30 \div 10 =$ .... **b)** $43 \div 10 =$ .... **c)** $5.8 \div 10 =$ ....

d) $63.2 \div 10 =$ .... **e)** $0.5 \div 10 =$ .... **f)** $400 \div 100 =$ ....

g) $423 \div 100 =$ .... **h)** $228.6 \div 100 =$ .... **i)** $61.5 \div 100 =$ ....

j) $2.96 \div 100 =$ .... **k)** $6000 \div 1000 =$ .... **l)** $6334 \div 1000 =$ ....

m) $753.6 \div 1000 =$ .... **n)** $8.15 \div 1000 =$ .... **o)** $80 \div 20 =$ ....

p) $860 \div 20 =$ .... **q)** $2400 \div 300 =$ .... **r)** $480 \div 40 =$ ....

s) $860 \div 200 =$ .... **t)** $63.9 \div 30 =$ ....

Q2 Ten people share a Lottery win of £62.
How much should each person receive?

....

Q3 Blackpool Tower is 158 m tall. A model of it is being built to a scale of 1 : 100.
How tall should the model be?

....

Q4 A box of 40 chocolates contains 56 g of saturated fat.
How much saturated fat is there in 1 chocolate?

....

Q5 Mark went on holiday to France. He exchanged £300 for 342 euros to spend while he was there. How many euros did he get for each £1?

....

Multiplying and Dividing Without a Calculator

There are lots of written methods you can use for multiplication and division. What you have to do is pick a method you like and practise using it on questions until you know it like the inside of your armour.

Q1 Use written methods to multiply the following:

no calculators!!

a) 23×2 = ...

b) 225×3 = ...

c) 546×5 = ...

d) 126×14 = ...

e) 413×26 = ...

f) 309×61 = ...

g) 847×53 = ...

Q2 Now use written methods to deal with these little blighters:

a) $834 \div 3$ = ...

b) $645 \div 5$ = ...

c) $702 \div 6$ = ...

d) $1000 \div 8$ = ...

e) $595 \div 17$ = ...

If the answer's not a whole number, give your answer as whole number plus remainder.

f) $768 \div 16$ = ...

g) $996 \div 24$ = ...

h) $665 \div 14$ = ...

Multiplying Decimals

A good way to multiply decimals is to ignore the decimal point to start with — just multiply the numbers. Then put the point back in and CHECK your answer looks sensible.

Do these calculations without using a calculator:

Q1 **a)** 3.2×4 **b)** 8.3×5 **c)** 6.4×3

= .... = ... = ...

d) 21×0.3 **e)** 63×0.23 **f)** 2.42×31

= .... = ... = ...

Q2 At a petrol station each pump shows a ready reckoner table.
Complete the table when the cost of unleaded petrol is 94.9p per litre.

Litres	Cost in pence
1	94.9
5	
10	
20	
50	

Q3 Jason can run 5.3 metres in 2 seconds.
How far will he run if he keeps up this pace for:

a) 20 seconds **b)** 60 seconds **c)** 5 minutes

....

Q4 Steve is ordering some new garden fencing online. He knows what size he wants in yards, feet and inches, but the website gives sizes in centimetres.
Using the information in the table, work out:

a) The number of centimetres in 3 inches.

....

b) The number of centimetres in 6 feet.

....

c) The number of centimetres in 45 yards.

....

Conversions

1 inch = 2.54 centimetres

1 foot = 12 inches

1 yard = 3 feet

1 metre = 3.28 feet

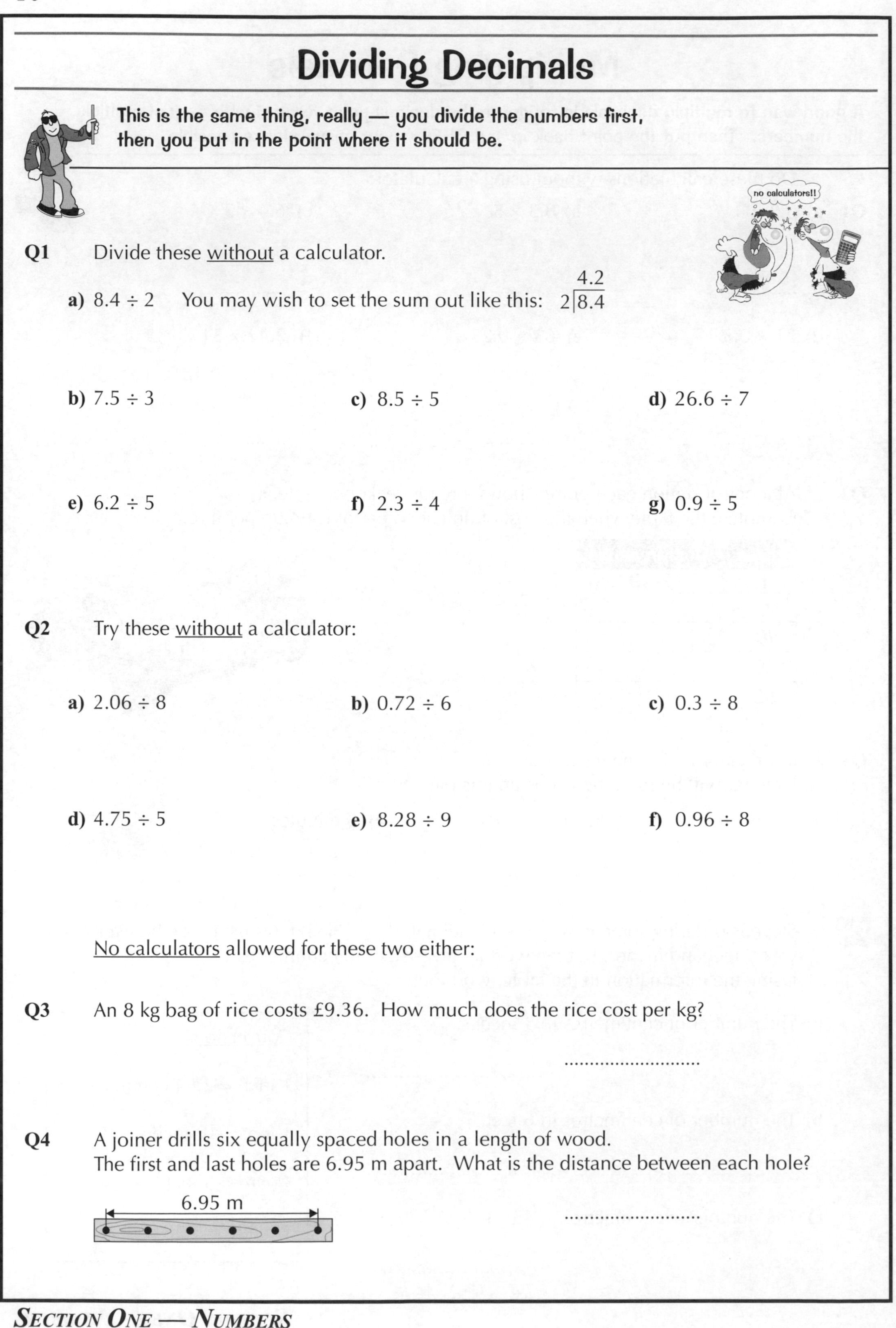

Dividing Decimals

This is the same thing, really — you divide the numbers first, then you put in the point where it should be.

Q1 Divide these without a calculator.

a) 8.4 ÷ 2 You may wish to set the sum out like this: $2\overline{)8.4}$ with 4.2 above

b) 7.5 ÷ 3 **c)** 8.5 ÷ 5 **d)** 26.6 ÷ 7

e) 6.2 ÷ 5 **f)** 2.3 ÷ 4 **g)** 0.9 ÷ 5

Q2 Try these without a calculator:

a) 2.06 ÷ 8 **b)** 0.72 ÷ 6 **c)** 0.3 ÷ 8

d) 4.75 ÷ 5 **e)** 8.28 ÷ 9 **f)** 0.96 ÷ 8

No calculators allowed for these two either:

Q3 An 8 kg bag of rice costs £9.36. How much does the rice cost per kg?

....

Q4 A joiner drills six equally spaced holes in a length of wood.
The first and last holes are 6.95 m apart. What is the distance between each hole?

6.95 m

....

Special Number Sequences

There are five special sequences: EVEN, ODD, SQUARE, CUBE and TRIANGLE NUMBERS. You really need to know them and their n^{th} terms.

EVEN SQUARE ODD CUBE TRIANGLE

Q1 What are these sequences called, and what are their next three terms?

a) 2, 4, 6, 8, ...

....

b) 1, 3, 5, 7, ...

....

c) 1, 4, 9, 16, ...

....

d) 1, 8, 27, 64, ...

....

e) 1, 3, 6, 10, ...

....

Q2 The following sequences are described in words. Write down their first four terms.

a) The prime numbers starting from 37.

....

b) The powers of 2 starting from 32.

....

c) The squares of odd numbers starting from $7^2 = 49$.

....

d) The powers of 10 starting from 1000.

....

e) The triangular numbers starting from 15.

The nth term is $\frac{1}{2}n(n + 1)$ — you're starting from n = 5.

....

Prime Numbers

Basically, prime numbers don't divide by anything (except 1 and themselves). 5 is prime — it will only divide by 1 or 5. 1 is an exception to this rule — it is not prime.

Q1 Write down the first ten prime numbers. ....

Q2 Give a reason for 27 not being a prime number. ....

Q3 Using any or all of the figures **1, 2, 3, 7** write down:

a) the smallest prime number ....

b) a prime number greater than 20 ....

c) a prime number between 10 and 20 ....

d) two prime numbers whose sum is 20 .... , ...

e) a number that is not prime ....

Q4 Find all the prime numbers between 40 and 50. ....

Q5 In the ten by ten square opposite, ring all the prime numbers.
The first three have been done for you.

1	(2)	(3)	4	(5)	6	7	8	9	10
11	12	13	14	15	16	17	18	19	20
21	22	23	24	25	26	27	28	29	30
31	32	33	34	35	36	37	38	39	40
41	42	43	44	45	46	47	48	49	50
51	52	53	54	55	56	57	58	59	60
61	62	63	64	65	66	67	68	69	70
71	72	73	74	75	76	77	78	79	80
81	82	83	84	85	86	87	88	89	90
91	92	93	94	95	96	97	98	99	100

Q6 A school ran three evening classes: judo, karate and kendo.
The judo class had 29 pupils, the karate class had 27 and the kendo class 23.
For which classes would the teacher have difficulty dividing the pupils into equal groups?

....

Q7 Find three sets of three prime numbers which add up to the following numbers:

10 ...., ..., ... **29** ...., ..., ... **41** ...., ..., ...

Multiples

The multiples of a number are its times table — if you need multiples of more than one number, do them separately then pick the ones in both lists.

Q1 What are the first five multiples of:

a) 4? ....

b) 7? ....

c) 12? ....

d) 18? ....

Q2 Find a number which is a multiple of:

A quick way to do these is just to multiply the numbers together.

a) 2 and 6 ....

b) 7 and 5 ....

c) 2 and 3 and 7 ....

d) 4 and 5 and 9 ....

Q3 Steven is making cheese straws for a dinner party. There will be **either** six or eight people at the party. He wants to be able to share the cheese straws out equally. How many cheese straws should he make?

There's more than one right answer to this question.

....

Q4 a) Find a number which is a multiple of 3 and 8 ....

b) Find another number which is a multiple of 3 and 8 ....

c) Find another number which is a multiple of 3 and 8 ....

Q5 Which of these numbers 14, 20, 22, 35, 50, 55, 70, 77, 99 are multiples of:

a) 2? ....

b) 5? ....

c) 7? ....

d) 11? ....

Factors

Factors multiply together to make other numbers.
E.g. $1 \times 6 = 6$ and $2 \times 3 = 6$, so 6 has factors 1, 2, 3 and 6.

Q1 List the factors of the following numbers.
Each factor should be written once, with no repeats.

a) 18 ....

b) 22 ....

c) 35 ....

d) 7 ....

e) 16 ....

f) 49 ....

g) 32 ....

h) 31 ....

i) 50 ....

j) 62 ....

k) 81 ....

l) 100 ....

Q2 a) I am a factor of 24.
I am an odd number.
I am greater than 1.
What number am I?

....

b) I am a factor of 30.
I am an even number.
I am less than 5.
What number am I?

....

Q3 Circle all the factors of 360 in this list of numbers.

1 2 3 4 5 6 7 8 9 10

Q4 48 students went on a geography field trip. Their teachers split them into equal groups.
Suggest five different ways that the teachers might have split up the students:

.... groups of ....

.... groups of ....

.... groups of ....

.... groups of ....

.... groups of ....

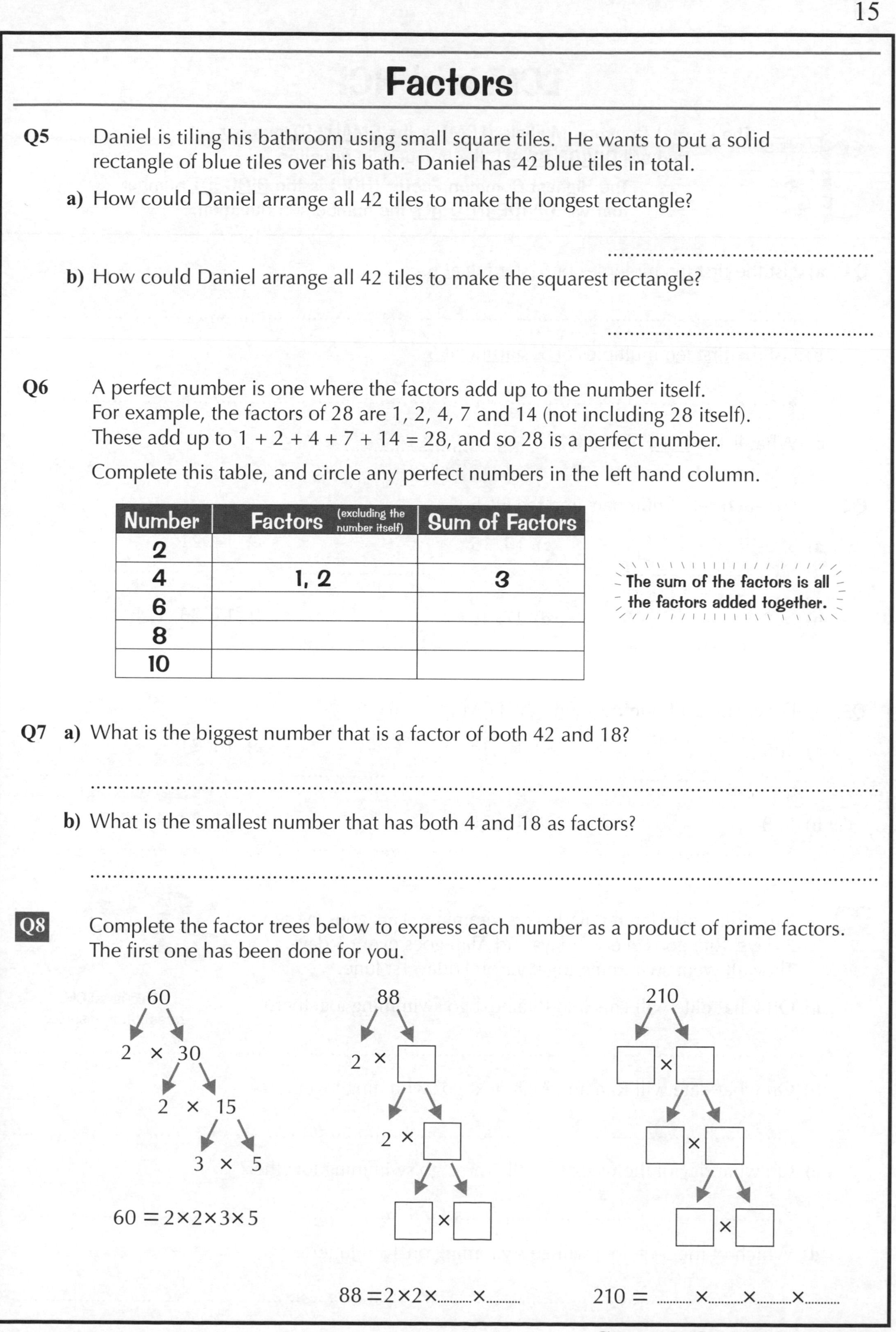

Factors

Q5 Daniel is tiling his bathroom using small square tiles. He wants to put a solid rectangle of blue tiles over his bath. Daniel has 42 blue tiles in total.

a) How could Daniel arrange all 42 tiles to make the longest rectangle?

....

b) How could Daniel arrange all 42 tiles to make the squarest rectangle?

....

Q6 A perfect number is one where the factors add up to the number itself.
For example, the factors of 28 are 1, 2, 4, 7 and 14 (not including 28 itself).
These add up to 1 + 2 + 4 + 7 + 14 = 28, and so 28 is a perfect number.

Complete this table, and circle any perfect numbers in the left hand column.

Number	Factors (excluding the number itself)	Sum of Factors
2		
4	1, 2	3
6		
8		
10		

The sum of the factors is all the factors added together.

Q7 a) What is the biggest number that is a factor of both 42 and 18?

....

b) What is the smallest number that has both 4 and 18 as factors?

....

Q8 Complete the factor trees below to express each number as a product of prime factors.
The first one has been done for you.

88 = 2×2×.........×.........

210 = ...×.........×.........×.........

LCM and HCF

Top tip

The Lowest Common Multiple (LCM) is the SMALLEST number that will DIVIDE BY ALL the numbers in question.

The Highest Common Factor (HCF) is the BIGGEST number that will DIVIDE INTO ALL the numbers in question.

Q1 **a)** List the first ten multiples of 6, starting at 6.

....

b) List the first ten multiples of 5, starting at 5.

....

c) What is the LCM of 5 and 6? ....

Q2 For each set of numbers find the HCF.

a) 3, 5 ....

b) 6, 8 ....

c) 10, 15 ....

d) 27, 48 ....

e) 14, 21 ....

f) 11, 33, 121 ....

Q3 For each set of numbers, find the LCM.

a) 3, 5 ....

b) 6, 8 ....

c) 10, 15 ....

d) 15, 18 ....

e) 14, 21 ....

f) 11, 33, 44 ....

Q4 Lars, Rita and Alan regularly go swimming. Lars goes every 2 days, Rita goes every 3 days and Alan goes every 5 days. They all went swimming together on Friday 1st June.

This is a LCM question in disguise.

a) On what date will Lars and Rita next go swimming together?

....

b) On what date will Rita and Alan next go swimming together?

....

c) On what day of the week will all 3 next go swimming together?

....

d) Which of the 3 (if any) will go swimming on 15th June?

....

Fractions, Decimals and Percentages

Q1 Change these fractions to decimals:

a) $\frac{1}{2}$ .... **b)** $\frac{3}{4}$ .... **c)** $\frac{7}{10}$ .... **d)** $\frac{19}{20}$ ....

e) $\frac{1}{100}$ .... **f)** $\frac{3}{8}$ .... **g)** $\frac{2}{1000}$ .... **h)** $\frac{1}{3}$ ....

Q2 Change these fractions to percentages:

a) $\frac{1}{4}$ .... **b)** $\frac{3}{10}$ .... **c)** $\frac{4}{5}$ .... **d)** $\frac{12}{25}$ ....

e) $\frac{8}{100}$ .... **f)** $\frac{2}{40}$ .... **g)** $\frac{7}{8}$ .... **h)** $\frac{11}{30}$ ....

Q3 Change these decimals to percentages:

a) 0.62 .... **b)** 0.74 .... **c)** 0.4 .... **d)** 0.9 ....

e) 0.07 .... **f)** 0.02 .... **g)** 0.125 .... **h)** 0.987 ....

Q4 Change these percentages to decimals:

a) 25% .... **b)** 49% .... **c)** 3% .... **d)** 30% ....

Q5 Change these percentages to fractions (in their lowest terms):

a) 75% .... **b)** 60% .... **c)** 15% .... **d)** 53% ....

Q6 Change these decimals to fractions (in their lowest terms):

a) 0.5 .... **b)** 0.8 .... **c)** 0.19 .... **d)** 0.25 ....

e) 0.64 .... **f)** 0.06 .... **g)** 0.125 .... **h)** 0.075 ....

A FRACTION IS A DECIMAL IS A PERCENTAGE —
they're all just different ways of saying "a bit of" something.

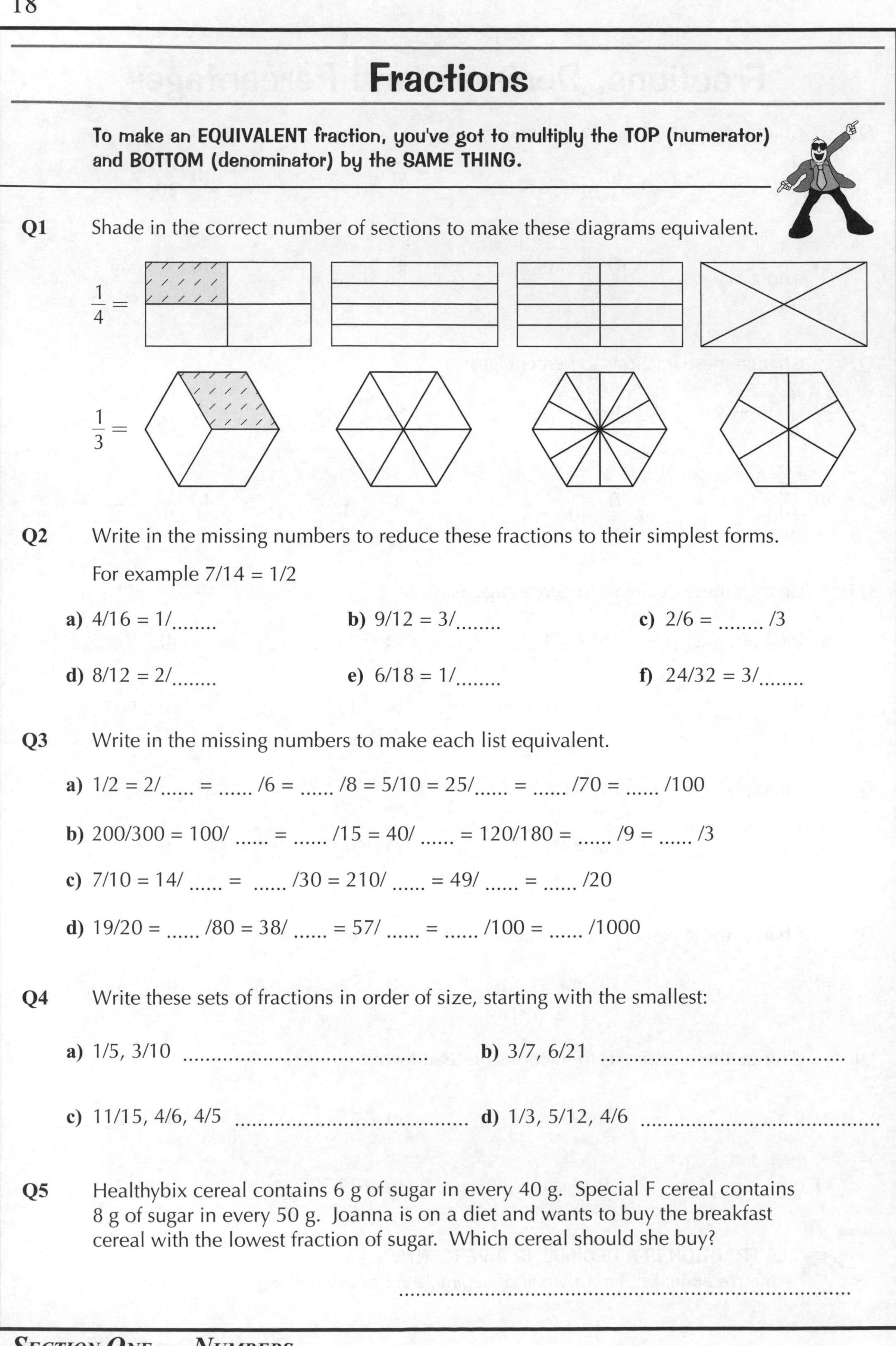

Fractions

To make an EQUIVALENT fraction, you've got to multiply the TOP (numerator) and BOTTOM (denominator) by the SAME THING.

Q1 Shade in the correct number of sections to make these diagrams equivalent.

$\frac{1}{4} =$

$\frac{1}{3} =$

Q2 Write in the missing numbers to reduce these fractions to their simplest forms.

For example 7/14 = 1/2

a) 4/16 = 1/........ **b)** 9/12 = 3/........ **c)** 2/6 = ... /3

d) 8/12 = 2/........ **e)** 6/18 = 1/........ **f)** 24/32 = 3/........

Q3 Write in the missing numbers to make each list equivalent.

a) 1/2 = 2/...... = ... /6 = ... /8 = 5/10 = 25/...... = ... /70 = ... /100

b) 200/300 = 100/ ...... = ... /15 = 40/ ...... = 120/180 = ... /9 = ... /3

c) 7/10 = 14/ ...... = ... /30 = 210/ ...... = 49/ ...... = ... /20

d) 19/20 = ... /80 = 38/ ...... = 57/ ...... = ... /100 = ... /1000

Q4 Write these sets of fractions in order of size, starting with the smallest:

a) 1/5, 3/10 .... **b)** 3/7, 6/21 ....

c) 11/15, 4/6, 4/5 .... **d)** 1/3, 5/12, 4/6 ....

Q5 Healthybix cereal contains 6 g of sugar in every 40 g. Special F cereal contains 8 g of sugar in every 50 g. Joanna is on a diet and wants to buy the breakfast cereal with the lowest fraction of sugar. Which cereal should she buy?

....

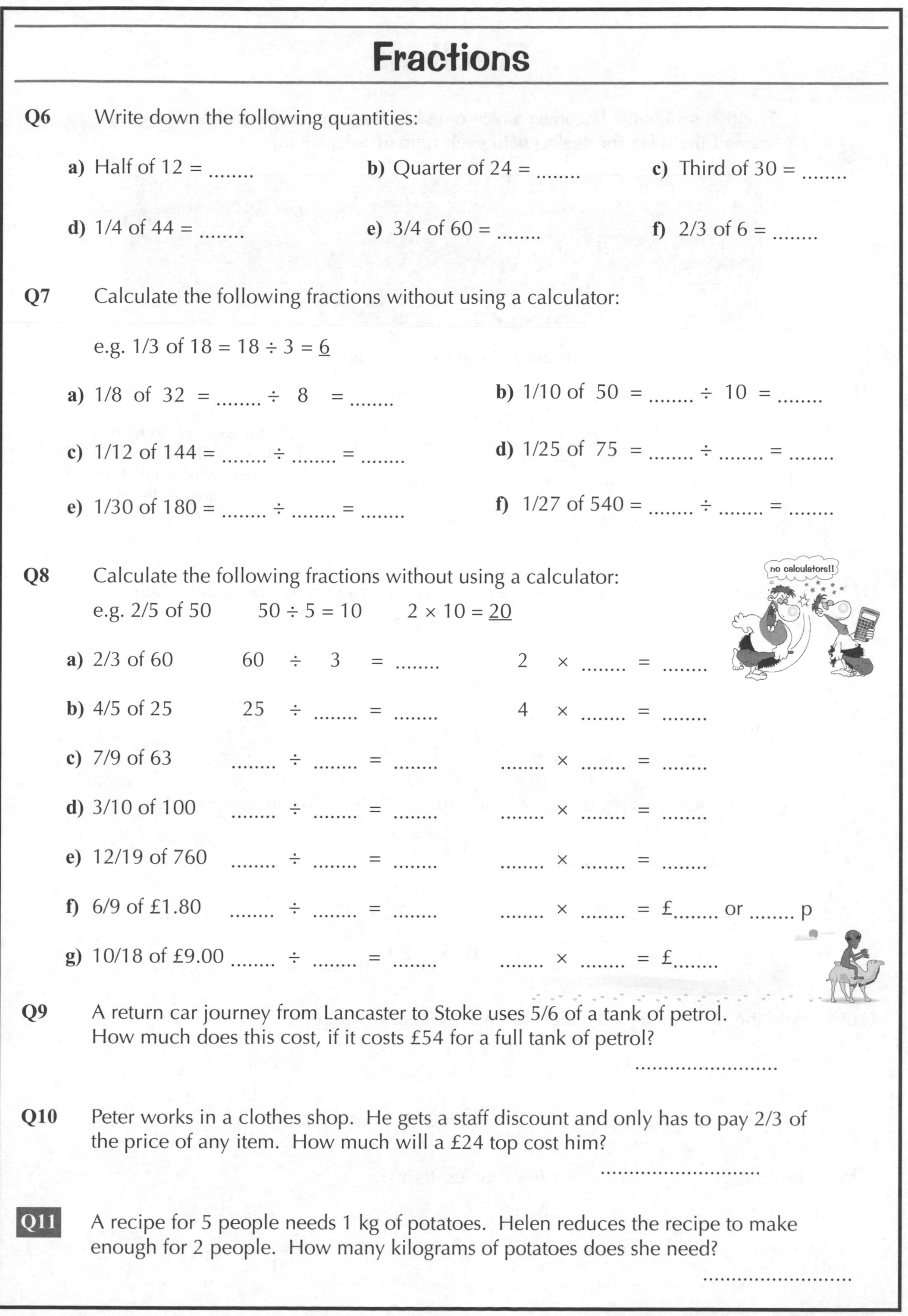

Fractions

Q6 Write down the following quantities:

a) Half of 12 = ... **b)** Quarter of 24 = ... **c)** Third of 30 = ...

d) 1/4 of 44 = ... **e)** 3/4 of 60 = ... **f)** 2/3 of 6 = ...

Q7 Calculate the following fractions without using a calculator:

e.g. 1/3 of 18 = 18 ÷ 3 = <u>6</u>

a) 1/8 of 32 = ... ÷ 8 = ...

b) 1/10 of 50 = ... ÷ 10 = ...

c) 1/12 of 144 = ... ÷ .... = ...

d) 1/25 of 75 = ... ÷ .... = ...

e) 1/30 of 180 = ... ÷ .... = ...

f) 1/27 of 540 = ... ÷ .... = ...

Q8 Calculate the following fractions without using a calculator:
e.g. 2/5 of 50 50 ÷ 5 = 10 2 × 10 = <u>20</u>

a) 2/3 of 60 60 ÷ 3 = ... 2 × .... = ...

b) 4/5 of 25 25 ÷ .... = ... 4 × .... = ...

c) 7/9 of 63 .... ÷ .... = ... × .... = ...

d) 3/10 of 100 .... ÷ .... = ... × .... = ...

e) 12/19 of 760 .... ÷ .... = ... × .... = ...

f) 6/9 of £1.80 .... ÷ .... = ... × .... = £........ or .... p

g) 10/18 of £9.00 .... ÷ .... = ... × .... = £........

Q9 A return car journey from Lancaster to Stoke uses 5/6 of a tank of petrol.
How much does this cost, if it costs £54 for a full tank of petrol?

....

Q10 Peter works in a clothes shop. He gets a staff discount and only has to pay 2/3 of the price of any item. How much will a £24 top cost him?

....

Q11 A recipe for 5 people needs 1 kg of potatoes. Helen reduces the recipe to make enough for 2 people. How many kilograms of potatoes does she need?

....

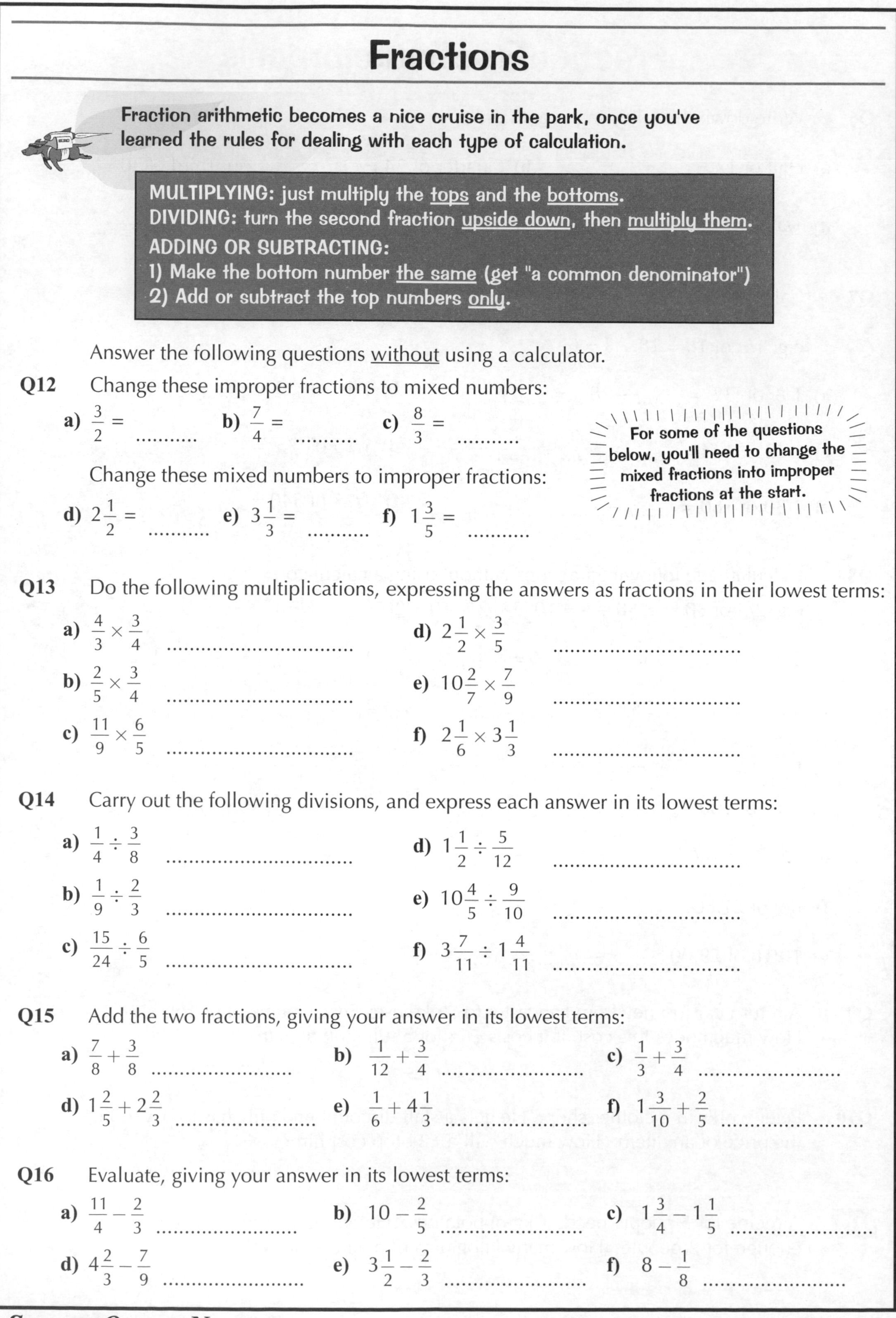

Fractions

Fraction arithmetic becomes a nice cruise in the park, once you've learned the rules for dealing with each type of calculation.

MULTIPLYING: just multiply the <u>tops</u> and the <u>bottoms</u>.
DIVIDING: turn the second fraction <u>upside down</u>, then <u>multiply them</u>.
ADDING OR SUBTRACTING:
1) Make the bottom number <u>the same</u> (get "a common denominator")
2) Add or subtract the top numbers <u>only</u>.

Answer the following questions <u>without</u> using a calculator.

Q12 Change these improper fractions to mixed numbers:

a) $\frac{3}{2}=$ .... **b)** $\frac{7}{4}=$ .... **c)** $\frac{8}{3}=$ ....

Change these mixed numbers to improper fractions:

d) $2\frac{1}{2}=$ .... **e)** $3\frac{1}{3}=$ .... **f)** $1\frac{3}{5}=$ ....

For some of the questions below, you'll need to change the mixed fractions into improper fractions at the start.

Q13 Do the following multiplications, expressing the answers as fractions in their lowest terms:

a) $\frac{4}{3}\times\frac{3}{4}$ .... **d)** $2\frac{1}{2}\times\frac{3}{5}$ ....

b) $\frac{2}{5}\times\frac{3}{4}$ .... **e)** $10\frac{2}{7}\times\frac{7}{9}$ ....

c) $\frac{11}{9}\times\frac{6}{5}$ .... **f)** $2\frac{1}{6}\times 3\frac{1}{3}$ ....

Q14 Carry out the following divisions, and express each answer in its lowest terms:

a) $\frac{1}{4}\div\frac{3}{8}$ .... **d)** $1\frac{1}{2}\div\frac{5}{12}$ ....

b) $\frac{1}{9}\div\frac{2}{3}$ .... **e)** $10\frac{4}{5}\div\frac{9}{10}$ ....

c) $\frac{15}{24}\div\frac{6}{5}$ .... **f)** $3\frac{7}{11}\div 1\frac{4}{11}$ ....

Q15 Add the two fractions, giving your answer in its lowest terms:

a) $\frac{7}{8}+\frac{3}{8}$ .... **b)** $\frac{1}{12}+\frac{3}{4}$ .... **c)** $\frac{1}{3}+\frac{3}{4}$ ....

d) $1\frac{2}{5}+2\frac{2}{3}$ .... **e)** $\frac{1}{6}+4\frac{1}{3}$ .... **f)** $1\frac{3}{10}+\frac{2}{5}$ ....

Q16 Evaluate, giving your answer in its lowest terms:

a) $\frac{11}{4}-\frac{2}{3}$ .... **b)** $10-\frac{2}{5}$ .... **c)** $1\frac{3}{4}-1\frac{1}{5}$ ....

d) $4\frac{2}{3}-\frac{7}{9}$ .... **e)** $3\frac{1}{2}-\frac{2}{3}$ .... **f)** $8-\frac{1}{8}$ ....

Fractions and Reciprocals

Try these questions without a calculator:

Q1 What fraction of 1 hour is:

a) 5 minutes? ....

b) 15 minutes? ....

c) 40 minutes? ....

Q2 If a TV programme lasts 40 minutes, what fraction of the programme is left after:

a) 10 minutes? ....

b) 15 minutes? ....

c) 35 minutes? ....

Q3 A café employs eighteen girls and twelve boys to wait at tables. Another six boys and nine girls work in the kitchen.
What fraction of the kitchen staff are girls?
What fraction of the employees are boys?

a) Fraction of kitchen staff who are girls ....

b) Fraction of employees who are boys ....

Use a calculator for the questions below.

Q4 If I pay my gas bill within seven days, I get a reduction of an eighth of the price. If my bill is £120, how much can I save? ....

Q5 Owen gives money to a charity directly from his wages, before tax. One sixth of his monthly earnings goes to the charity, then one fifth of what's left gets deducted as tax. How much money goes to charity and on tax, and how much is left, if he earns £2400?

Charity ...., Tax ...., Left over ....

Q6 Write down the reciprocals of the following values.
Leave your answers as whole numbers or fractions.

a) 7 **b)** 12 **c)** $\frac{3}{8}$ **d)** $-\frac{1}{2}$

Q7 Use your calculator to work out the reciprocals of the following values.
Write your answers as whole numbers or decimals.

a) 12 **b)** $\sqrt{2}$ **c)** π **d)** 0.008

Don't get put off by all the padding in the questions — you've just got to pick out the important stuff.

Fractions and Recurring Decimals

This is about as scary as fractions get, but don't panic!
It's still just about learning the rules.

All fractions can be written as TERMINATING or RECURRING decimals

If the denominator (bottom number) of a fraction has PRIME FACTORS of only 2 and/or 5, then it's a TERMINATING DECIMAL.
If the denominator (bottom number) of a fraction has OTHER PRIME FACTORS, then it's a RECURRING DECIMAL.

Answer the following questions without using a calculator.

Q1 Express each number as a product of its prime factors:

a) 6 .... **d)** 15 ....

b) 20 .... **e)** 9 ....

c) 55 .... **f)** 32 ....

Q2 Say whether each of these fractions will give a terminating or recurring decimal:

a) 1/6 .... **d)** 2/15 ....

b) 1/20 .... **e)** 4/9 ....

c) 3/55 .... **f)** 7/32 ....

no calculators!!

To convert a recurring decimal to a fraction, just remember this rule:
The fraction always has the repeating unit on the top and the same number of nines on the bottom.

Q3 Write the following decimals as fractions in their lowest form:

a) 0.222... .... **b)** 0.444... .... **c)** 0.808080... ....

d) $0.\dot{3}$ .... **e)** $0.\dot{6}$ .... **f)** $0.\overline{16}$ ....

Ratios

Ratios compare quantities of the same kind — so if the units aren't mentioned, they've got to be the same in each bit of the ratio.

Q1 What is the ratio in each of these pictures?

a)

b)

Circles to triangles .... to ....

Triangles to circles .... : ....

Big stars to small stars .... : ....

Q2 Write each of these ratios in its simplest form. The first one is done for you.

a) 4 to 6

2 : 3

b) 15 to 21

...... : ......

c) 14 to 42

...... : ......

d) 72 to 45

...... : ......

e) 24 cm to 36 cm

...... : ......

f) 350 g to 2 kg

...... : ......

g) 42p to £1.36

...... : ......

Watch out for ones like f) and g) — you need to make the units the same first.

Q3 I have some friends coming round for dinner and want to cook my favourite fish pie. My recipe serves 4 people, but I will need enough pie for 9 people.

a) How many potatoes will I need to cook my fish pie for everyone?

....

b) How much haddock will I need?

....

c) If eggs come in boxes of 6, how many boxes should I buy?

....

....

....

RECIPE FOR MY FAVOURITE FISH PIE

400 g haddock
8 large potatoes
1 tin mushy peas
12 eggs
1 pinch of salt
2 tablespoons curry powder

Ratios

I think I can spot a Golden Rule lurking here...
DIVIDE FOR ONE, THEN TIMES FOR ALL.

Q4 To make grey paint, black and white paint are mixed in the ratio 5:3.
How much black paint would be needed with:

a) 6 litres of white? ....

b) 12 litres of white? ....

c) 21 litres of white? ....

Q5 To make salad dressing you mix olive oil and vinegar in the ratio 5:2.
How much olive oil is needed with:

a) 10 ml of vinegar? ....

b) 30 ml of vinegar? ....

c) 42 ml of vinegar? ....

Q6 Sarah works as a waitress. Each week, she splits her wage into spending money and savings in the ratio 7:3.

a) One week, Sarah earns £130.
How much should she put in her savings that week? ....

b) The next week, Sarah put £42 into her savings.
How much did she earn in total that week? ....

Q7 The ratio of men to women at a football match was 11:4.
How many men were there if there were:

a) 2000 women? ....

b) 8460 women? ....

Ratios

Q8 Divide the following quantities in the given ratio.

For example:

£400 in the ratio 1 : 4 $\quad$ 1 + 4 = 5 $\quad$ £400 ÷ 5 = £80

1 × £80 = £80 and 4 × £80 = £320 $\quad$ £80 : £320

a) 100 g in the ratio 1 : 4 $\quad$ .... + ... = ... $\quad$ .... ÷ .... = ...

.... × .... = ... and .... × .... = ...

= .... : ....

b) 500 m in the ratio 2 : 3 = ... : ....

c) £12 000 in the ratio 1 : 2 = ... : ....

d) 6.3 kg in the ratio 3 : 4 = ... : ....

e) £8.10 in the ratio 4 : 5 = ... : ....

Q9 Now try these...

a) Adam and Mags win £24 000. They split the money in the ratio 1 : 5. How much does Adam get?

....

b) Sunil and Paul work in a restaurant. Any tips they earn are split in the ratio 3 : 4. One night they earned £28 in tips between them. Who got the most tip money? How much did they get?

.... got £.......................

c) The total distance covered in a triathlon (swimming, cycling and running) is 15 km. It is split in the ratio 2 : 3 : 5. How far is each section?

Swimming = ..., cycling = ..., running = ...

A great way to check your answer works is to add up the individual quantities — they should add up to the original amount.

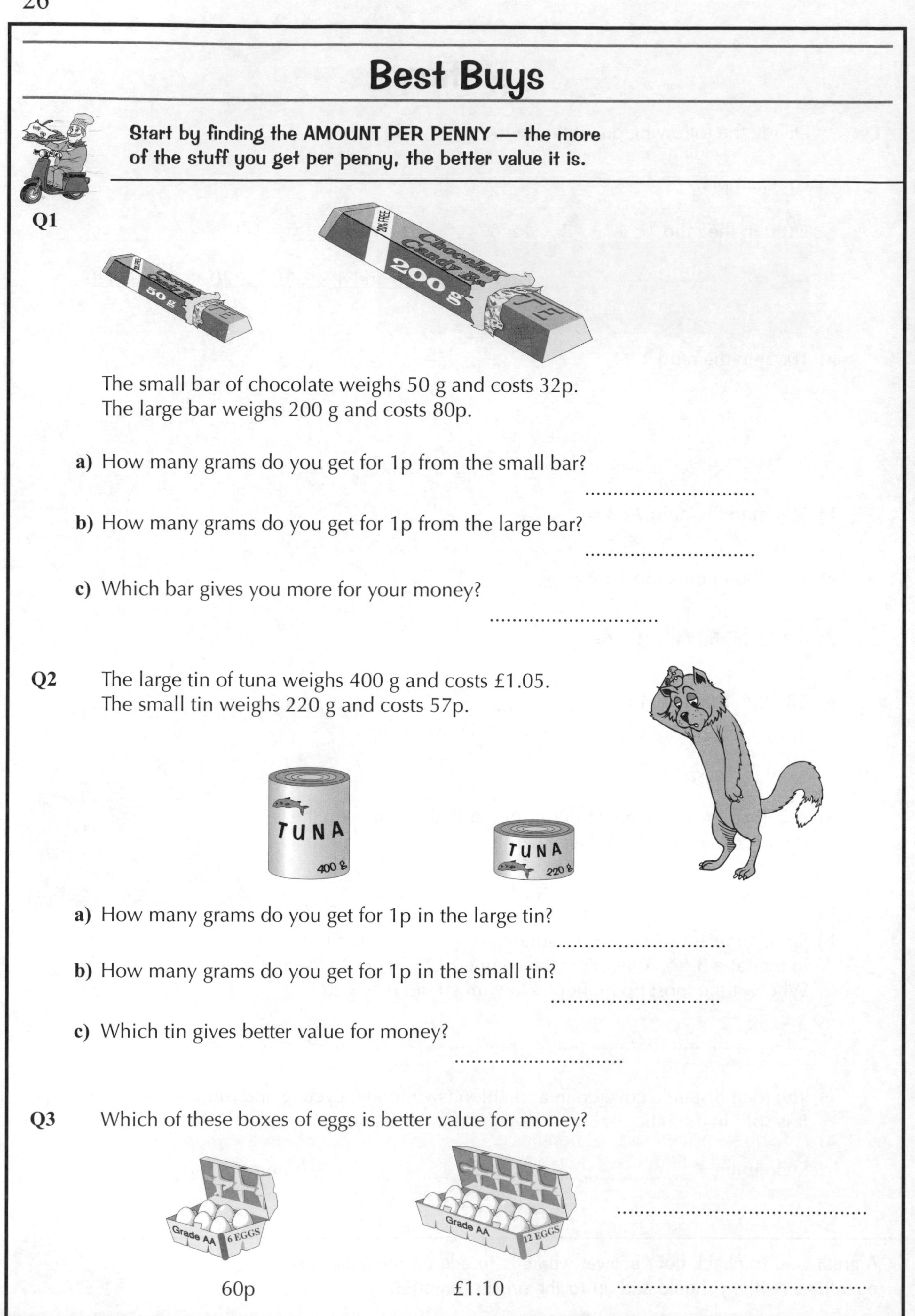

Best Buys

Start by finding the AMOUNT PER PENNY — the more of the stuff you get per penny, the better value it is.

Q1

The small bar of chocolate weighs 50 g and costs 32p.
The large bar weighs 200 g and costs 80p.

a) How many grams do you get for 1p from the small bar?

....

b) How many grams do you get for 1p from the large bar?

....

c) Which bar gives you more for your money?

....

Q2 The large tin of tuna weighs 400 g and costs £1.05.
The small tin weighs 220 g and costs 57p.

a) How many grams do you get for 1p in the large tin?

....

b) How many grams do you get for 1p in the small tin?

....

c) Which tin gives better value for money?

....

Q3 Which of these boxes of eggs is better value for money?

60p £1.10

....

....

Percentages

Finding "something %" of "something-else" is really quite simple — so you'd better be sure you know how.

1) "OF" means "×".
2) "PER CENT" means "OUT OF 100".

Example: 30% of 150 would be translated as $\frac{30}{100} \times 150 = 45$.

no calculators!!

Q1 Without using a calculator, work out the following percentages:

a) 50% of £12 = ... **b)** 25% of £20 = ...

c) 10% of £50 .... **d)** 5% of £50 = ...

e) 30% of £50 = ... **f)** 75% of £80 = ...

g) 10% of 90 cm = ... **h)** 10% of 4.39 kg = ...

Now have a go at these using your calculator:

i) 8% of £16 = ...

j) 85% of 740 kg = ...

k) 40% of 40 minutes = ...

A school has 750 pupils.

l) If 56% of the pupils are boys, what percentage are girls? ....

m) How many boys are there in the school? ....

n) One day, 6% of the pupils were absent. How many pupils was this? ....

o) 54% of the pupils have a school lunch, 38% bring sandwiches and the rest go home for lunch. How many pupils go home for lunch?

....

Q2 Work out these percentage questions:

a) A T-shirt is priced at £18. Ted has a 15% off voucher to use in the shop. How much will the T-shirt cost Ted if he uses his voucher?

....

b) A savings account gains 2.5% interest per year. How much interest will a saver earn if they have £200 in their account for a year?

....

Percentages

Q3 John bought a new PC. The tag in the shop said it cost £890 + VAT.
If VAT is charged at $17\frac{1}{2}\%$, how much did he pay?

....

Q4 Admission to Wonder World is £18 for adults. A child ticket is 60% of the adult price.

a) How much will it cost for one adult and 4 children to enter Wonder World?

....

b) How much will two adults and three children spend on entrance tickets?

....

Q5 Daphne earns an annual wage of £18 900. She doesn't pay tax on the first £6400 that she earns. How much income tax does she pay if the rate of tax is:

a) 20% ? .... **b)** 40% ? ....

Q6 The cost of a train ticket changes with inflation. If the annual rate of inflation is 2% and my train ticket home cost £28 last year, how much will my ticket home cost this year?

....

Q7 A double-glazing salesman is paid 10% commission on every sale he makes.
In addition he is paid a £50 bonus if the sale is over £500.

a) If a customer buys £499 worth of windows from the salesman, what is his commission?

....

b) How much extra will he earn if he persuades the customer in **a)** to spend an extra £20?

....

c) How much does he earn if he sells £820 worth of windows?

....

Q8

Bed & breakfast £37 per person. Evening meal £15 per person.

THE PICKLED PARROT

2 people stay at The Pickled Parrot for 2 nights and have an evening meal each night. How much is the total cost, if VAT is added at $17\frac{1}{2}\%$?

....

Percentages

Divide the new amount by the old, then × 100... or if you've been practising on your calc, you'll know you can just press the % button for the 2nd bit...

Q9 Express each of the following as a percentage. Round off if necessary.

a) £6 of £12 = ...

b) £4 of £16 = ...

c) 600 kg of 750 kg = ...

d) 6 hours of one day = ...

e) 1 month of a year = ...

f) 25 m of 65 m = ...

Q10 Calculate the percentage saving of the following:

e.g. Trainers: Was £72 Now £56 Saved ..£16.. of £72 = ...22...%

a) Jeans: Was £45 Now £35 Saved .... of £45 = ...%

b) CD: Was £14.99 Now £12.99 Saved .... of .... = ...%

c) Shirt: Was £27.50 Now £22.75 Saved .... of .... = ...%

d) TV: Was £695 Now £435 Saved .... = ...%

e) Microwave: Was £132 Now £99 Saved .... = ...%

Q11 a) If Andy Mint won 3 out of the 5 sets in the Wimbledon Men's Final, what percentage of the sets did he not win?

....%

b) Of a resort's 25 000 tourists last Summer, 23 750 were between 16 and 30 years old. What percentage of the tourists were not in this age group?

....%

c) Jeff went on a diet. At the start he weighed 92 kg, after one month he weighed 84 kg. What is his percentage weight loss?

....%

d) In their first game of the season, Nowcastle had 24 567 fans watching the game. By the final game there were 32 741 fans watching. What is the percentage increase in the number of fans?

....%

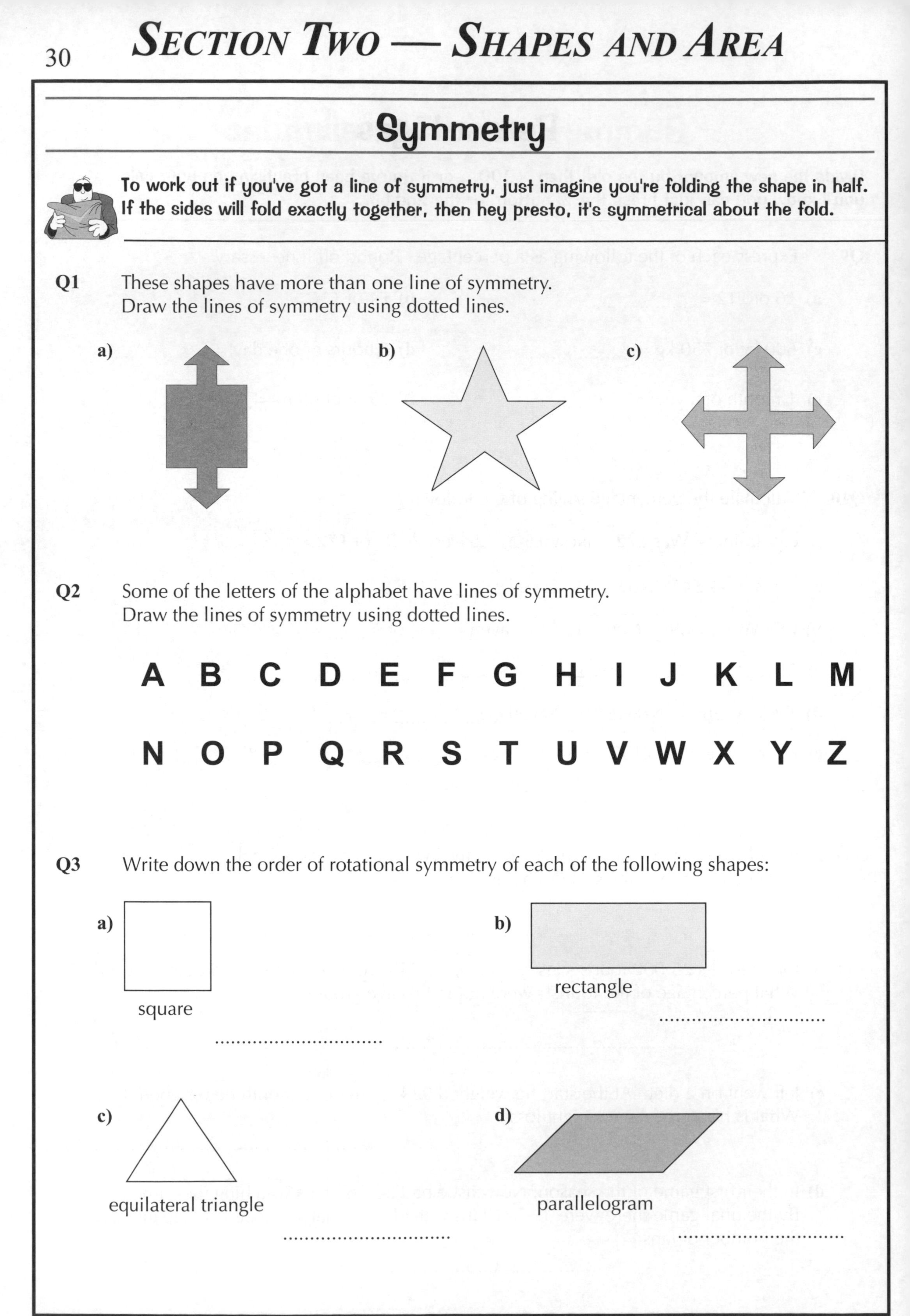

Symmetry

To work out if you've got a line of symmetry, just imagine you're folding the shape in half. If the sides will fold exactly together, then hey presto, it's symmetrical about the fold.

Q1 These shapes have more than one line of symmetry.
Draw the lines of symmetry using dotted lines.

a) **b)** **c)**

Q2 Some of the letters of the alphabet have lines of symmetry.
Draw the lines of symmetry using dotted lines.

A B C D E F G H I J K L M

N O P Q R S T U V W X Y Z

Q3 Write down the order of rotational symmetry of each of the following shapes:

a) square ....

b) rectangle ....

c) equilateral triangle ....

d) parallelogram ....

Symmetry and Tessellations

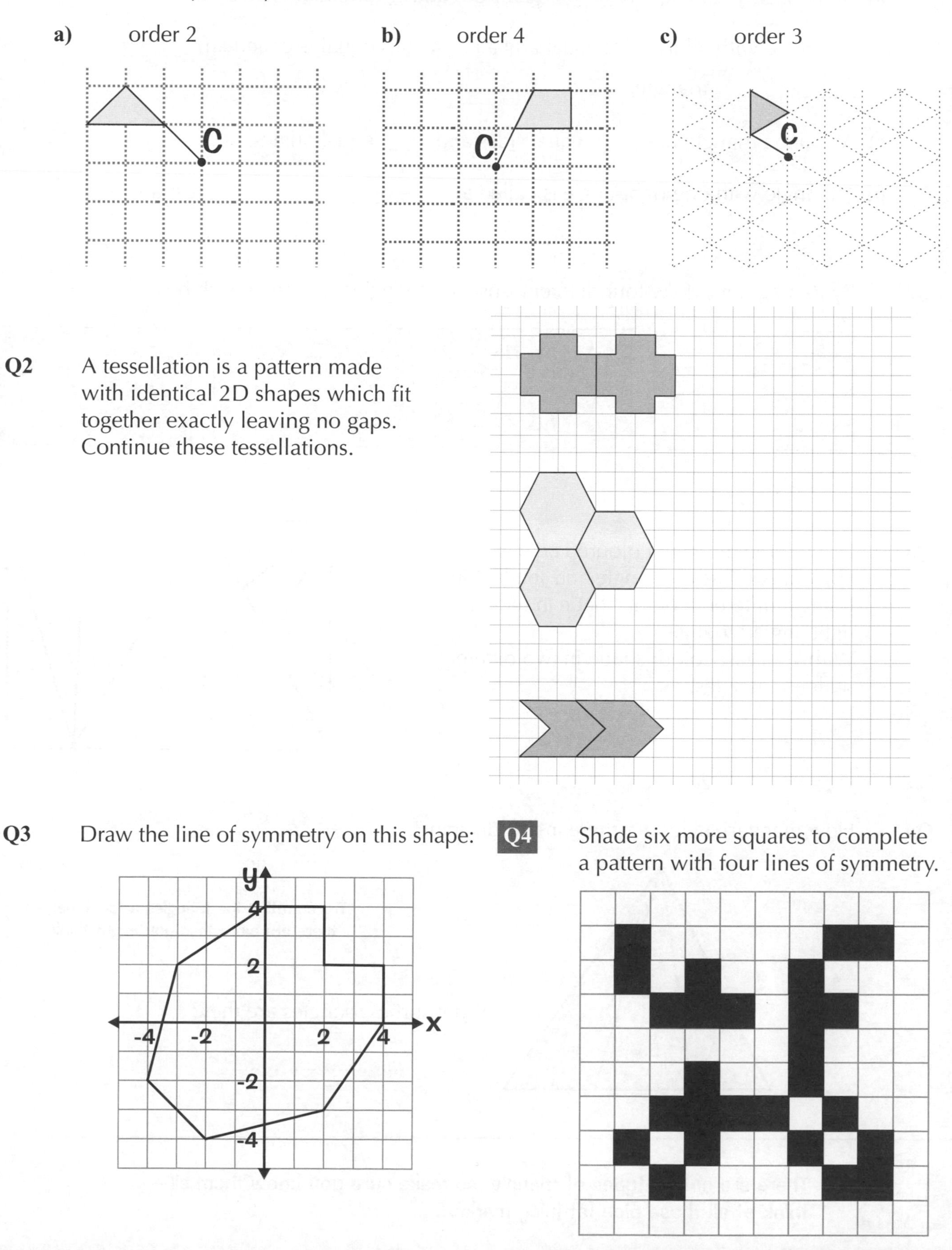

Q1 Complete the following diagrams so that they have rotational symmetry about centre C of the order stated:

a) order 2 **b)** order 4 **c)** order 3

Q2 A tessellation is a pattern made with identical 2D shapes which fit together exactly leaving no gaps. Continue these tessellations.

Q3 Draw the line of symmetry on this shape:

Q4 Shade six more squares to complete a pattern with four lines of symmetry.

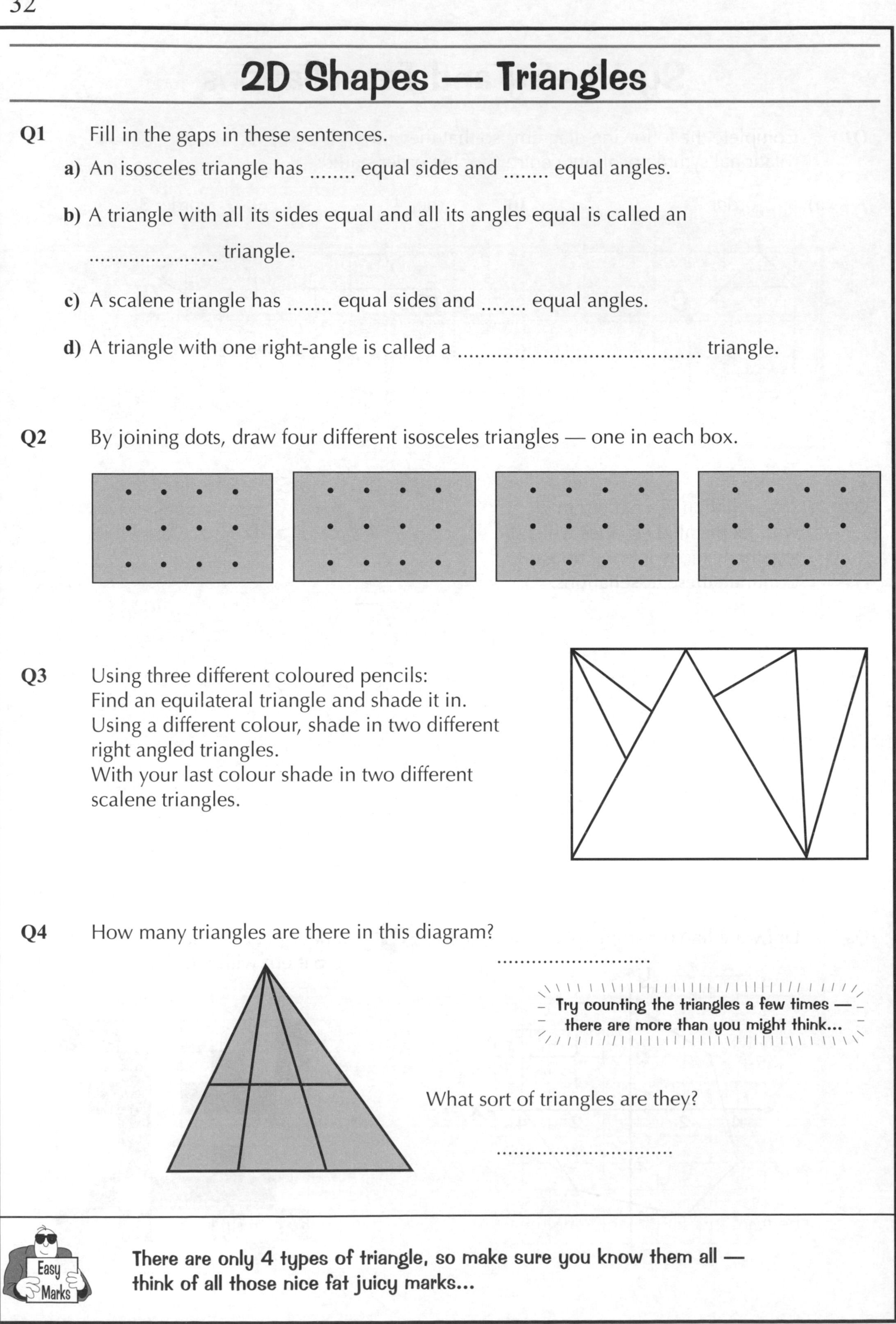

2D Shapes — Triangles

Q1 Fill in the gaps in these sentences.

a) An isosceles triangle has .... equal sides and .... equal angles.

b) A triangle with all its sides equal and all its angles equal is called an .... triangle.

c) A scalene triangle has .... equal sides and .... equal angles.

d) A triangle with one right-angle is called a .... triangle.

Q2 By joining dots, draw four different isosceles triangles — one in each box.

Q3 Using three different coloured pencils:
Find an equilateral triangle and shade it in.
Using a different colour, shade in two different right angled triangles.
With your last colour shade in two different scalene triangles.

Q4 How many triangles are there in this diagram?

....

Try counting the triangles a few times — there are more than you might think...

What sort of triangles are they?

....

There are only 4 types of triangle, so make sure you know them all — think of all those nice fat juicy marks...

2D Shapes — Quadrilaterals

Here's a few easy marks for you — all you've got to do is remember the shapes and a few facts about them... it's a waste of marks not to bother.

Q1 Fill in the blanks in the table.

NAME	DRAWING	DESCRIPTION
Square		Sides of equal length. Opposite sides parallel. Four right angles.
....	....	Opposite sides parallel and the same length. Four right angles.
....		Opposite sides are .... and equal. Opposite angles are equal.
Trapezium		Only .... sides are parallel.
Rhombus		A parallelogram but with all sides ....
Kite	....	Two pairs of adjacent equal sides.

Q2 Stephen is measuring the angles inside a parallelogram for his maths homework. To save time, he measures just one and works out what the other angles must be from this. If the angle he measures is 52°, what are the other three?

.........., ..., ...,

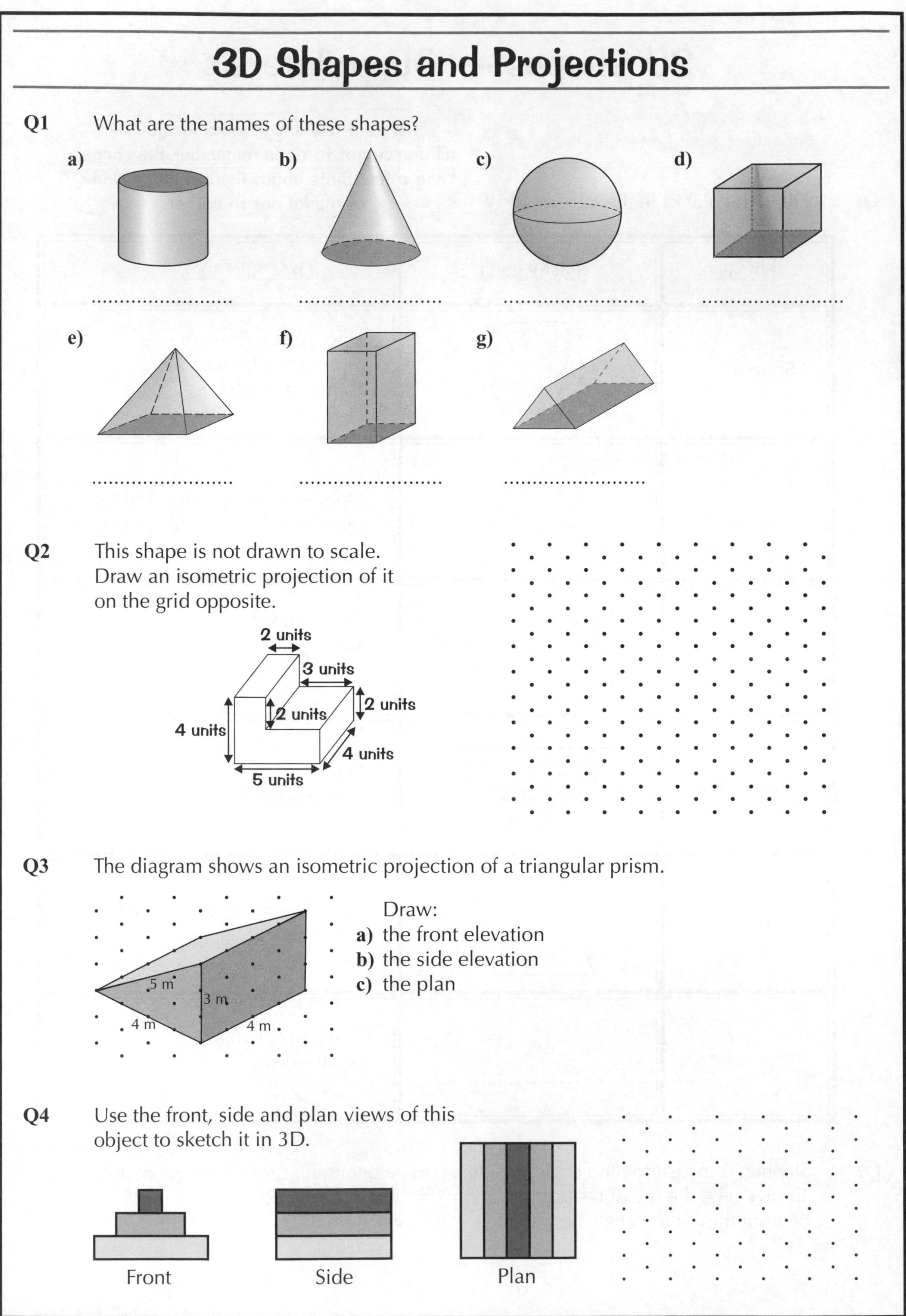

3D Shapes and Projections

Q1 What are the names of these shapes?

a) ....

b) ....

c) ....

d) ....

e) ....

f) ....

g) ....

Q2 This shape is not drawn to scale. Draw an isometric projection of it on the grid opposite.

Q3 The diagram shows an isometric projection of a triangular prism.

Draw:

a) the front elevation

b) the side elevation

c) the plan

Q4 Use the front, side and plan views of this object to sketch it in 3D.

Regular Polygons

Regular polygons are just shapes that follow certain rules — which makes them ideal exam question material...

Q1 Describe what a regular polygon is.

....

....

Q2 Sketch a regular hexagon and draw in all its lines of symmetry.
State the order of rotational symmetry.

Rotational symmetry is just the number of positions in which the shape looks the same.

Order of rotational symmetry is ....

Q3 Complete the following table:

Name	Sides	Lines of Symmetry	Order of Rotational Symmetry
Equilateral Triangle			
Square		4	
Regular Pentagon			
Regular Hexagon	6		
Regular Heptagon	7		
Regular Octagon			8
Regular Decagon	10		

Q4 The diagram shows a regular nonagon.
What is the size of angle a?

a

40°

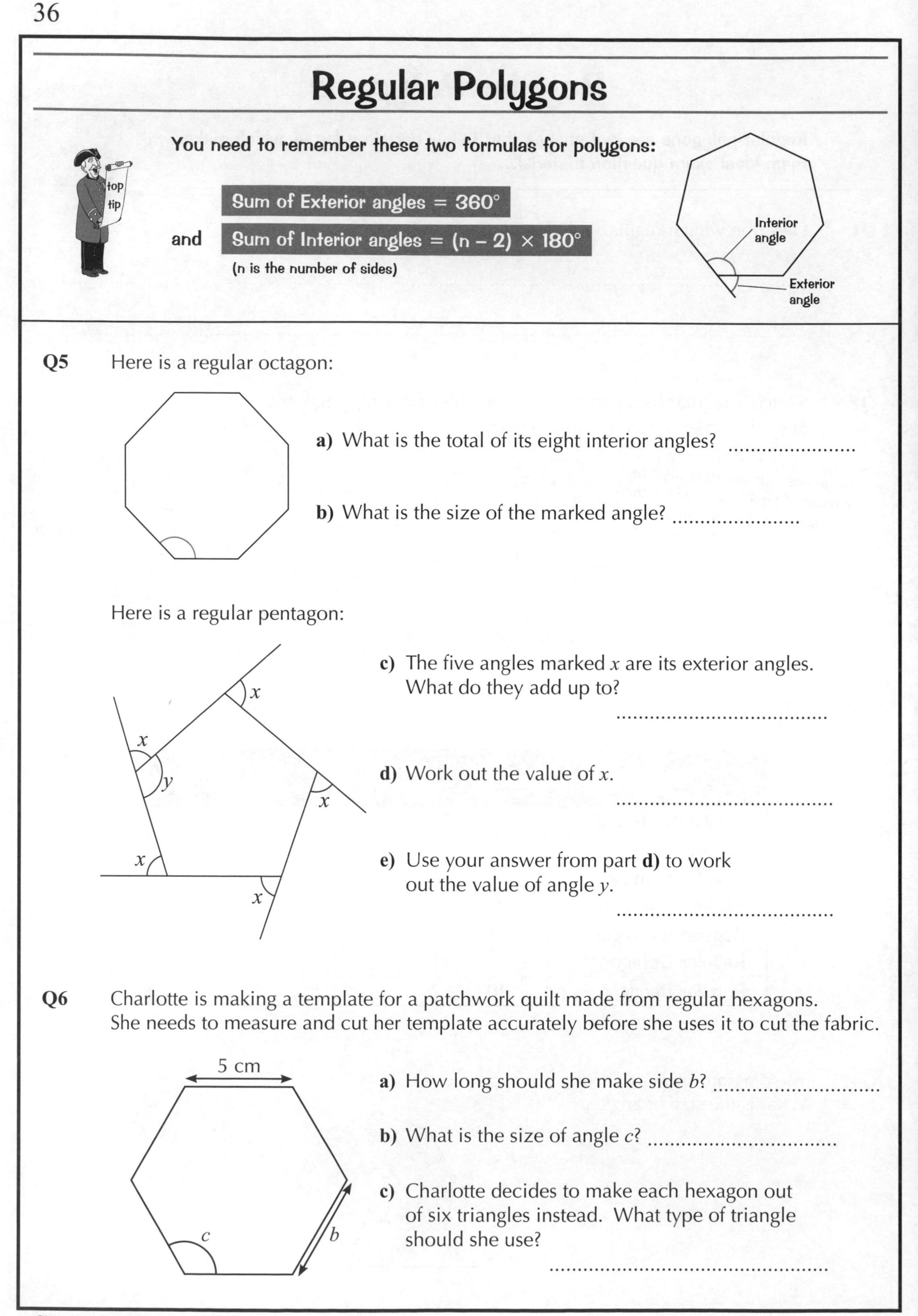

Regular Polygons

You need to remember these two formulas for polygons:

top tip

Sum of Exterior angles = 360°

and **Sum of Interior angles = (n – 2) × 180°**

(n is the number of sides)

Q5 Here is a regular octagon:

a) What is the total of its eight interior angles? ....

b) What is the size of the marked angle? ....

Here is a regular pentagon:

c) The five angles marked x are its exterior angles. What do they add up to?

....

d) Work out the value of x.

....

e) Use your answer from part **d)** to work out the value of angle y.

....

Q6 Charlotte is making a template for a patchwork quilt made from regular hexagons. She needs to measure and cut her template accurately before she uses it to cut the fabric.

a) How long should she make side b? ....

b) What is the size of angle c? ....

c) Charlotte decides to make each hexagon out of six triangles instead. What type of triangle should she use?

....

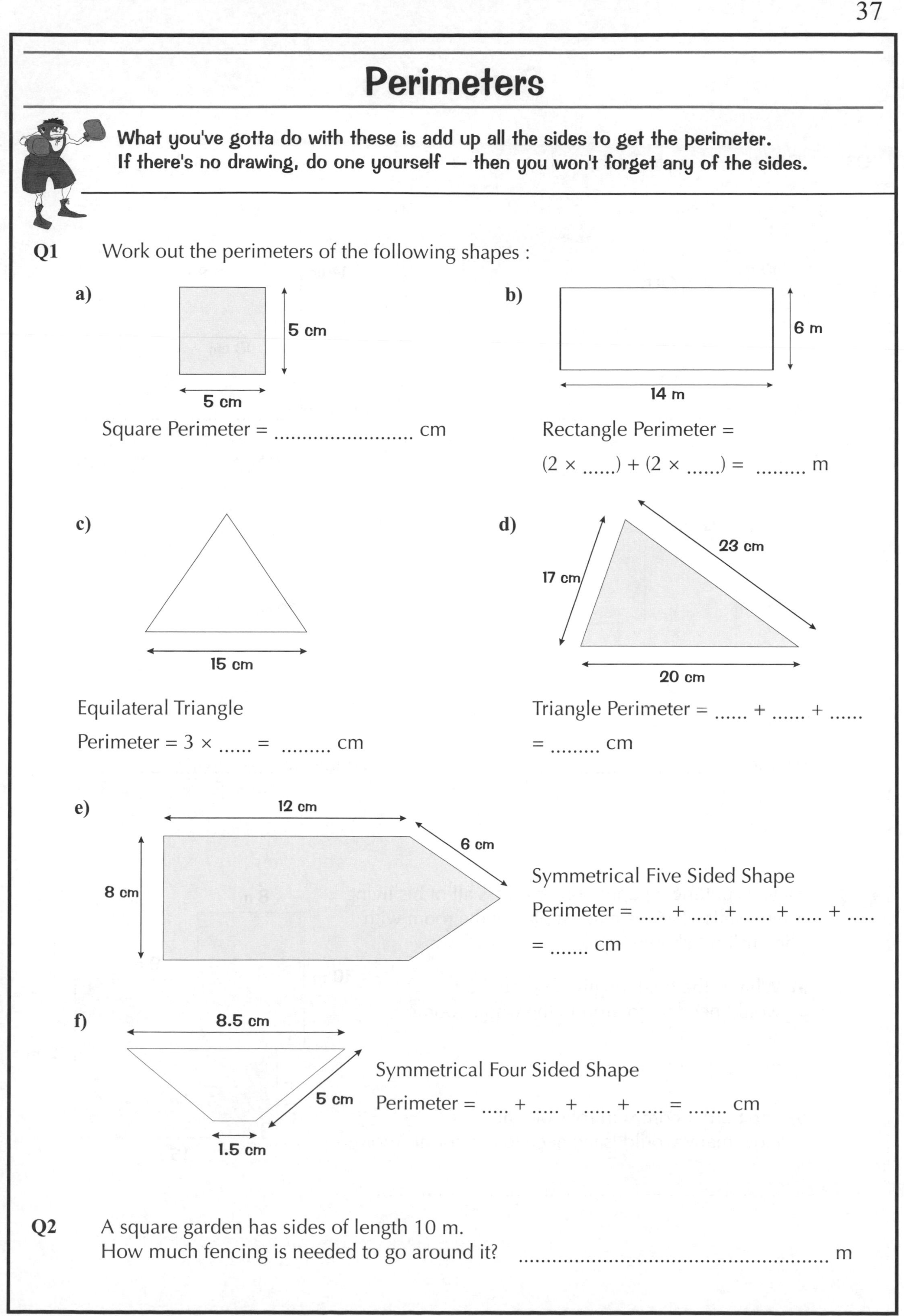

Perimeters

**What you've gotta do with these is add up all the sides to get the perimeter.
If there's no drawing, do one yourself — then you won't forget any of the sides.**

Q1 Work out the perimeters of the following shapes :

a) 5 cm, 5 cm

Square Perimeter = ... cm

b) 6 m, 14 m

Rectangle Perimeter =

$(2 \times ......) + (2 \times ......) =$ .... m

c) 15 cm

Equilateral Triangle

Perimeter = $3 \times$ ...... = ... cm

d) 17 cm, 23 cm, 20 cm

Triangle Perimeter = ... + ... + ...

= .... cm

e) 12 cm, 6 cm, 8 cm

Symmetrical Five Sided Shape

Perimeter = ... + ... + ... + ... + ...

= .... cm

f) 8.5 cm, 5 cm, 1.5 cm

Symmetrical Four Sided Shape

Perimeter = ... + ... + ... + ... = ... cm

Q2 A square garden has sides of length 10 m.
How much fencing is needed to go around it? .... m

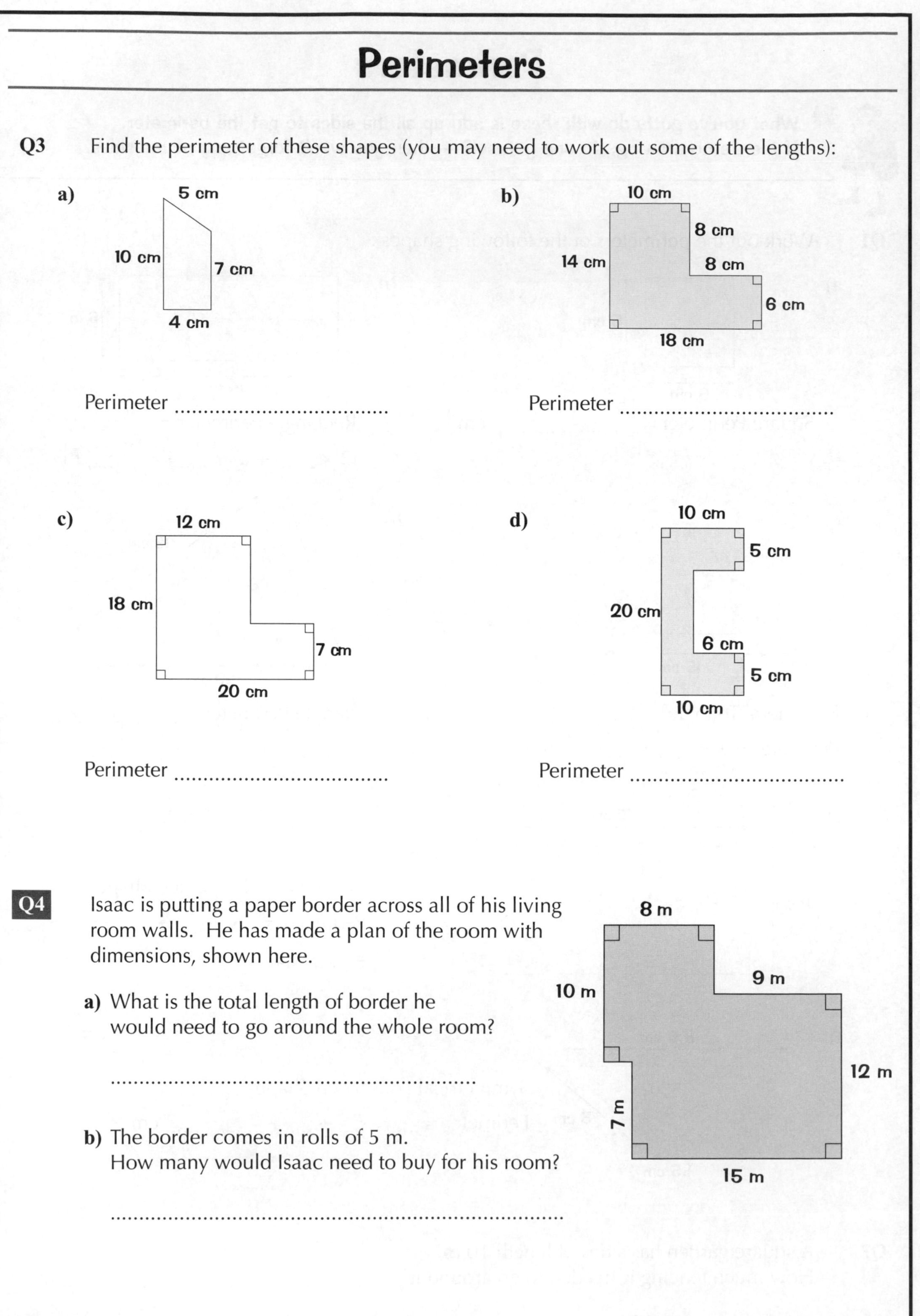

Perimeters

Q3 Find the perimeter of these shapes (you may need to work out some of the lengths):

a)

Perimeter ....

b)

Perimeter ....

c)

Perimeter ....

d)

Perimeter ....

Q4 Isaac is putting a paper border across all of his living room walls. He has made a plan of the room with dimensions, shown here.

a) What is the total length of border he would need to go around the whole room?

....

b) The border comes in rolls of 5 m.
How many would Isaac need to buy for his room?

....

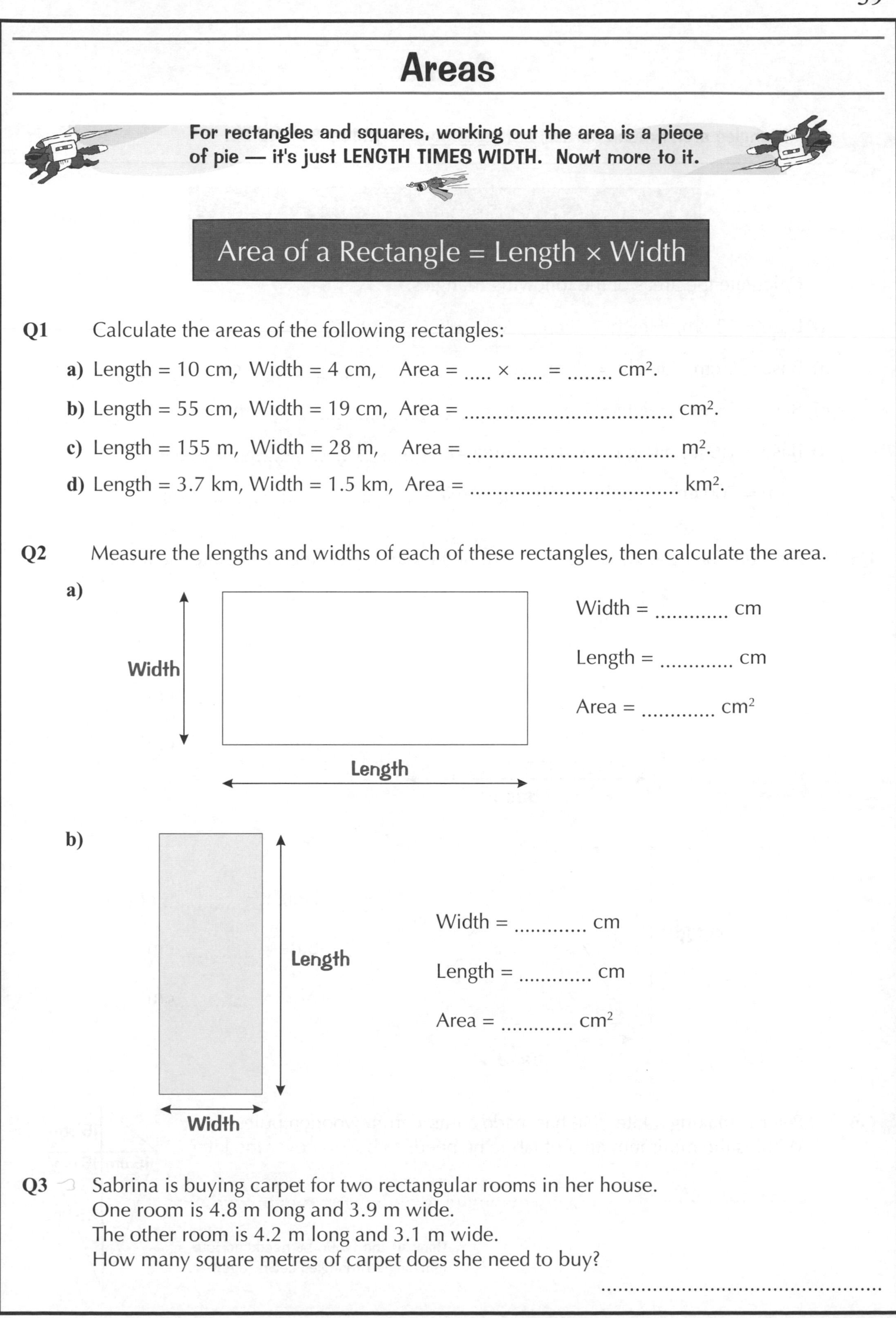

Areas

For rectangles and squares, working out the area is a piece of pie — it's just LENGTH TIMES WIDTH. Nowt more to it.

Area of a Rectangle = Length × Width

Q1 Calculate the areas of the following rectangles:

a) Length = 10 cm, Width = 4 cm, Area = ... × ..... = ... cm².

b) Length = 55 cm, Width = 19 cm, Area = ... cm².

c) Length = 155 m, Width = 28 m, Area = ... m².

d) Length = 3.7 km, Width = 1.5 km, Area = ... km².

Q2 Measure the lengths and widths of each of these rectangles, then calculate the area.

a)

Width = ... cm

Length = ... cm

Area = ... cm²

b)

Width = ... cm

Length = ... cm

Area = ... cm²

Q3 Sabrina is buying carpet for two rectangular rooms in her house.
One room is 4.8 m long and 3.9 m wide.
The other room is 4.2 m long and 3.1 m wide.
How many square metres of carpet does she need to buy?

....

Areas

Triangles aren't much harder, but remember to TIMES BY THE ½.

Area of a Triangle = ½ (Base × Height)

Q4 Calculate the areas of the following triangles:

a) Base = 12 cm, Height = 9 cm, Area = ½ (....... × ....) = ... cm^2.

b) Base = 5 cm, Height = 3 cm, Area = ... cm^2.

c) Base = 25 m, Height = 7 m, Area = ... m^2.

d) Base = 1.6 m, Height = 6.4 m, Area = ... m^2.

e) Base = 700 cm, Height = 350 cm, Area = ... cm^2.

Q5 Measure the base and height of each of these triangles, then calculate the area.

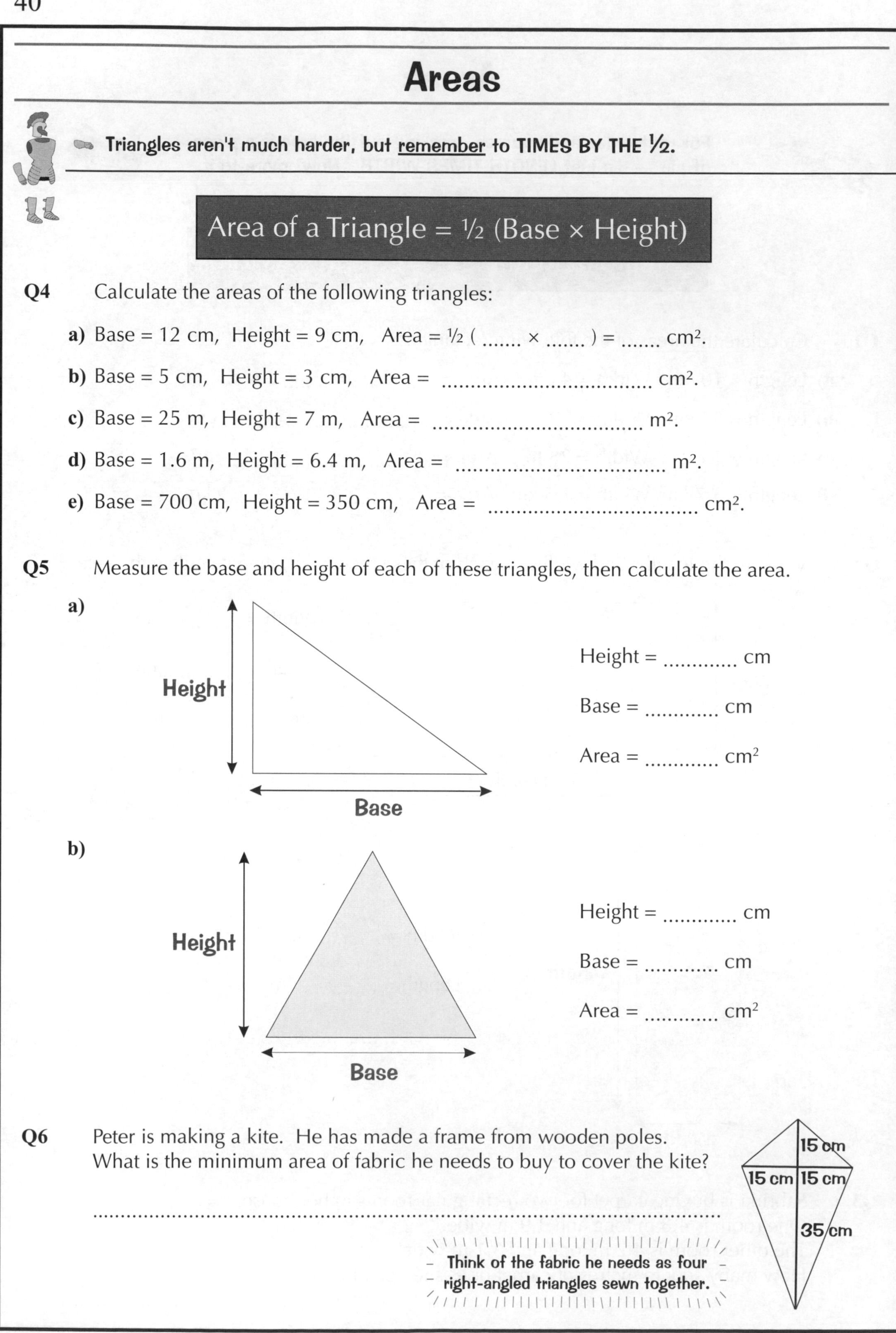

a)

Height = ... cm

Base = ... cm

Area = ... cm^2

b)

Height = ... cm

Base = ... cm

Area = ... cm^2

Q6 Peter is making a kite. He has made a frame from wooden poles.
What is the minimum area of fabric he needs to buy to cover the kite?

....

Think of the fabric he needs as four right-angled triangles sewn together.

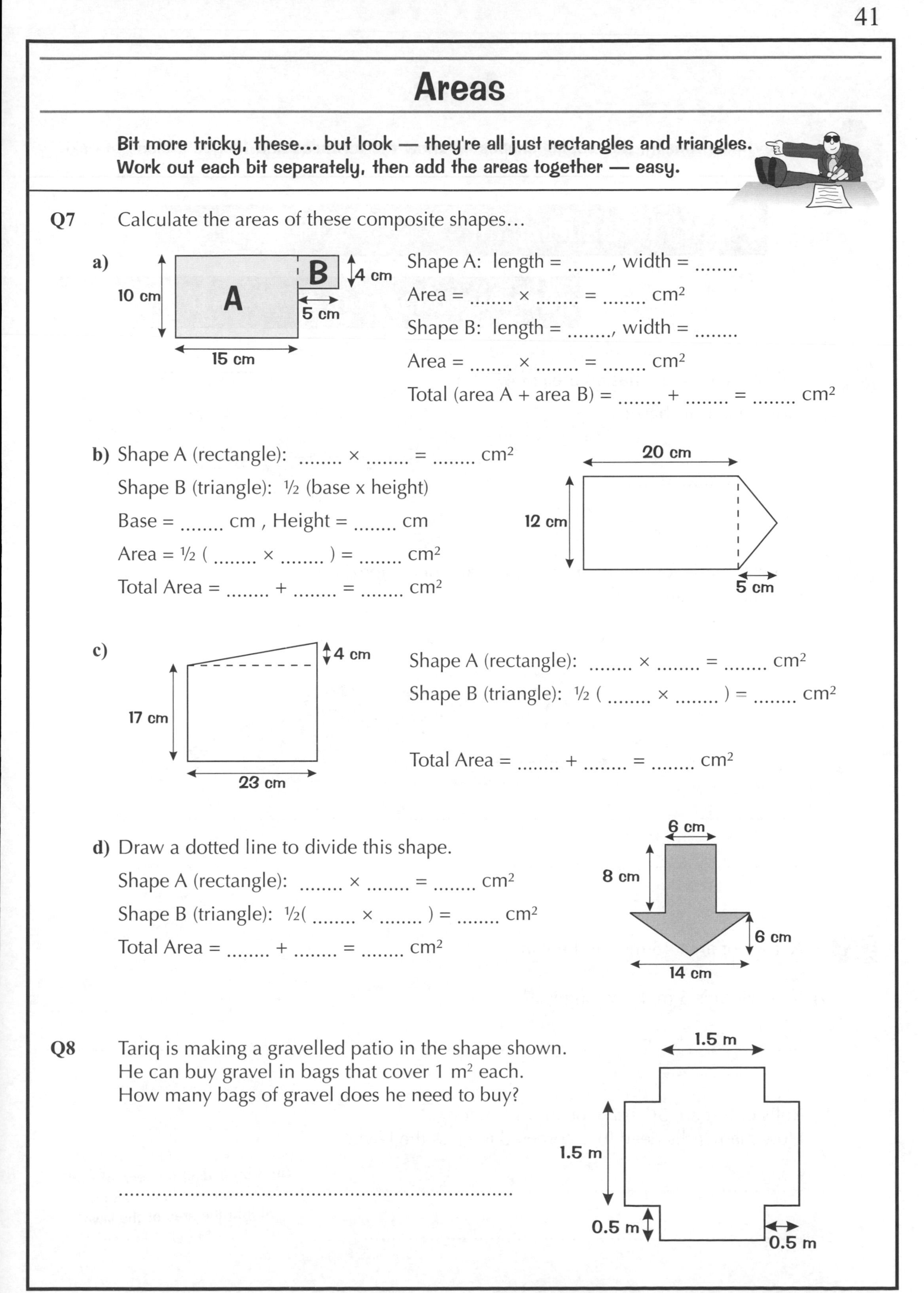

Areas

Bit more tricky, these... but look — they're all just rectangles and triangles. Work out each bit separately, then add the areas together — easy.

Q7 Calculate the areas of these composite shapes…

a)

Shape A: length = ..., width = ...

Area = ... × .... = ... cm^2

Shape B: length = ..., width = ...

Area = ... × .... = ... cm^2

Total (area A + area B) = ... + ... = ... cm^2

b) Shape A (rectangle): .... × .... = ... cm^2

Shape B (triangle): ½ (base x height)

Base = ... cm , Height = ... cm

Area = ½ (........ × ....) = ... cm^2

Total Area = ... + ... = ... cm^2

c)

Shape A (rectangle): .... × .... = ... cm^2

Shape B (triangle): ½ (........ × ....) = ... cm^2

Total Area = ... + ... = ... cm^2

d) Draw a dotted line to divide this shape.

Shape A (rectangle): .... × .... = ... cm^2

Shape B (triangle): ½(........ × ....) = ... cm^2

Total Area = ... + ... = ... cm^2

Q8 Tariq is making a gravelled patio in the shape shown. He can buy gravel in bags that cover 1 m^2 each. How many bags of gravel does he need to buy?

....

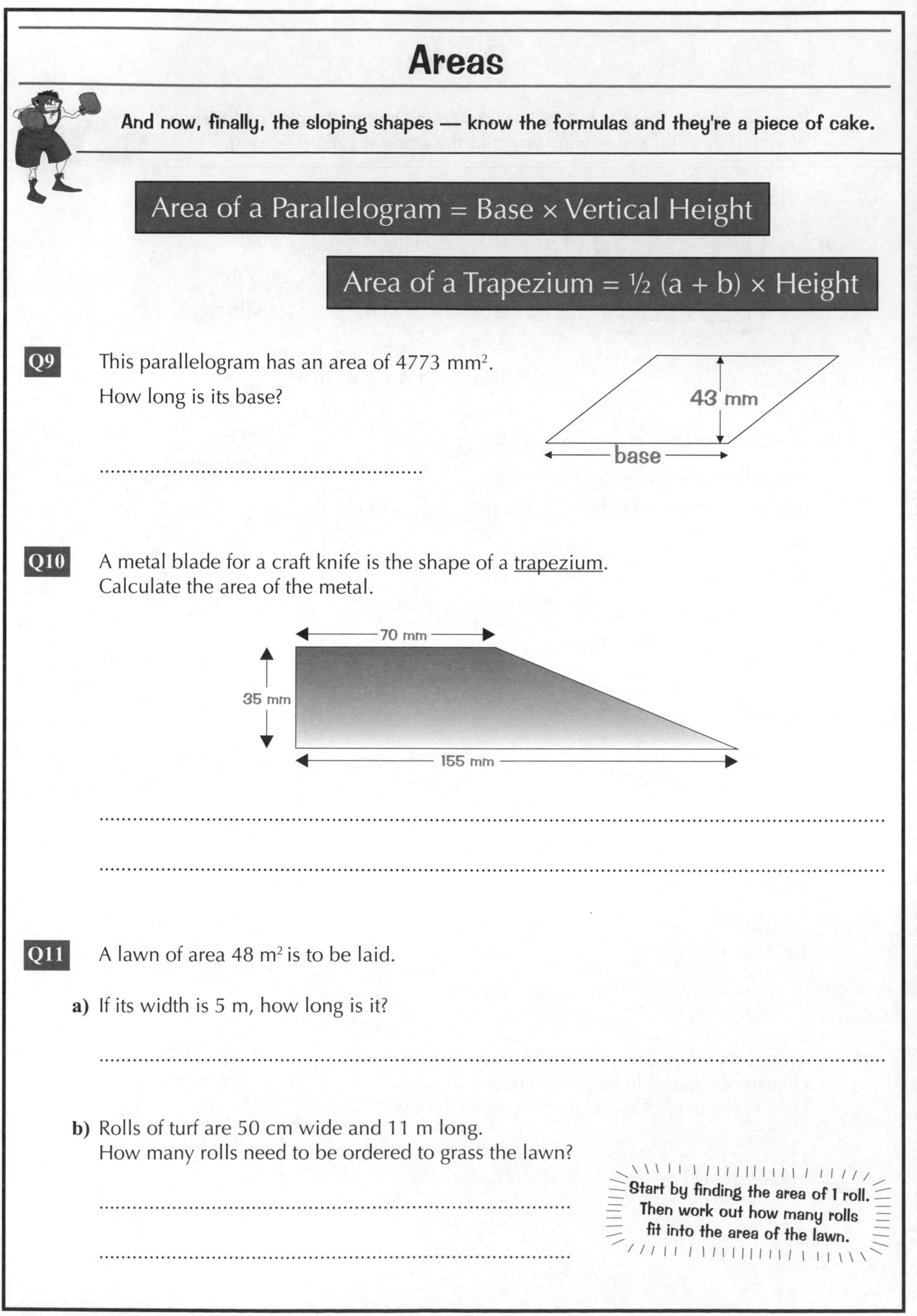

Areas

And now, finally, the sloping shapes — know the formulas and they're a piece of cake.

Area of a Parallelogram = Base × Vertical Height

Area of a Trapezium = ½ (a + b) × Height

Q9 This parallelogram has an area of 4773 mm².
How long is its base?

43 mm
base

....

Q10 A metal blade for a craft knife is the shape of a <u>trapezium</u>.
Calculate the area of the metal.

70 mm
35 mm
155 mm

....

....

Q11 A lawn of area 48 m² is to be laid.

a) If its width is 5 m, how long is it?

....

b) Rolls of turf are 50 cm wide and 11 m long.
How many rolls need to be ordered to grass the lawn?

....

....

Start by finding the area of 1 roll. Then work out how many rolls fit into the area of the lawn.

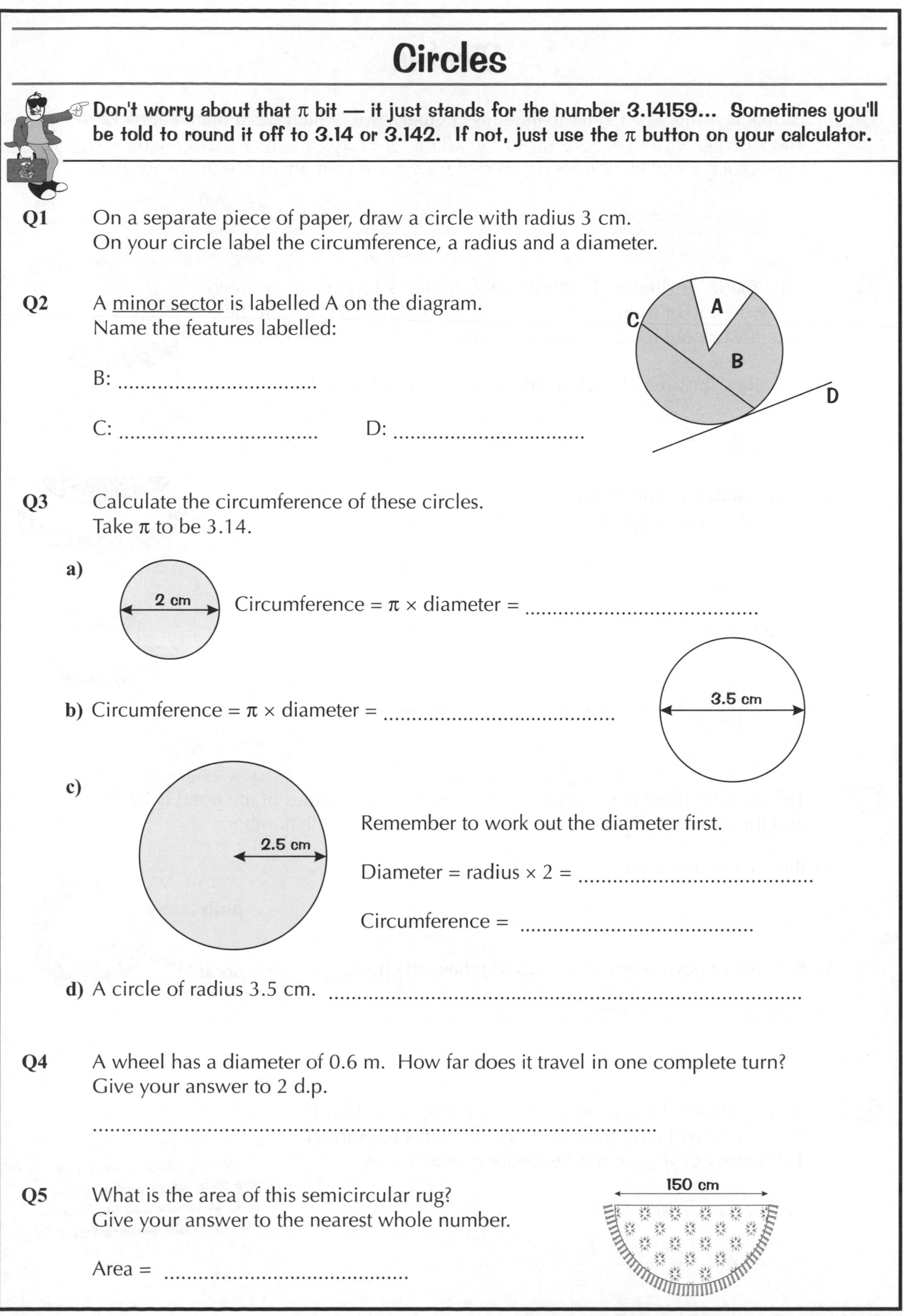

Circles

Don't worry about that π bit — it just stands for the number 3.14159... Sometimes you'll be told to round it off to 3.14 or 3.142. If not, just use the π button on your calculator.

Q1 On a separate piece of paper, draw a circle with radius 3 cm.
On your circle label the circumference, a radius and a diameter.

Q2 A minor sector is labelled A on the diagram.
Name the features labelled:

B: ....

C: .... D: ....

Q3 Calculate the circumference of these circles.
Take π to be 3.14.

a)

Circumference = π × diameter = ...

b) Circumference = π × diameter = ...

c)

Remember to work out the diameter first.

Diameter = radius × 2 = ...

Circumference = ...

d) A circle of radius 3.5 cm. ....

Q4 A wheel has a diameter of 0.6 m. How far does it travel in one complete turn?
Give your answer to 2 d.p.

....

Q5 What is the area of this semicircular rug?
Give your answer to the nearest whole number.

Area = ...

Circles

Q6 Patrick is using weed killer to treat his circular lawn, which has a diameter of 12 m. The weed killer instructions say to use 20 ml for every square metre of lawn to be treated. How much weed killer does he need to use? Give your answer to the nearest ml.

....

Q7 Anja made a cylindrical birthday cake of radius 15 cm. She wanted to ice the top of each quarter in a different colour. This meant she had to work out the area of each quarter of the cake to be iced.

a) Calculate the area of each quarter of the cake to be iced.

....

b) Anja wanted to cut the cake in half and put a ribbon around one half. How long would the ribbon have to be?

....

c) Anja only had 50 cm of ribbon. Would it be enough to put around a quarter of her cake?

....

Q8 This circular pond has a circular path around it. The radius of the pond is 72 m and the path is 2 m wide. Calculate to the nearest whole number:

a) the area of the pond.

....

path

pond

b) the area of paving stones needed to repave the path.

....

Q9 A bicycle wheel has a radius of 32 cm. Calculate how far the bicycle will have travelled after the wheel has turned 100 times. Give your answer to the nearest metre.

....

....

You only need to work out the distance covered by one of the wheels, as the bike will travel the same distance.

Volume

Finding volumes of cubes and cuboids is just like finding areas of squares and rectangles — except you've got an extra side to multiply by.

Q1 A match box measures 7 cm by 4 cm by 5 cm.
What is its volume?

....

Q2 What is the volume of a cube of side:

a) 5 cm? ....

b) 9 cm? ....

c) 15 cm? ....

Q3 Sam has a rectangular jelly mould measuring 16 cm by 16 cm by 6 cm.
Is the mould big enough to hold 1600 cm^3 of jelly?

....

Q4 A box measures 9 cm by 5 cm by 8 cm.

a) What is its volume? ....

b) What is the volume of a box twice
as long, twice as wide and twice as tall? ....

Q5 Which holds more, a box measuring 12 cm by 5 cm by 8 cm or a box measuring
10 cm by 6 cm by 9 cm?

....

Q6 A rectangular swimming pool is 12 m wide and 18 m long.
How many cubic metres of water are needed to fill the pool to a depth of 2 m?

....

Q7 An ice cube measures 2 cm by 2 cm by 2 cm. What is its volume? ....
Is there enough room in a container measuring
8 cm by 12 cm by 10 cm for 100 ice cubes? ....

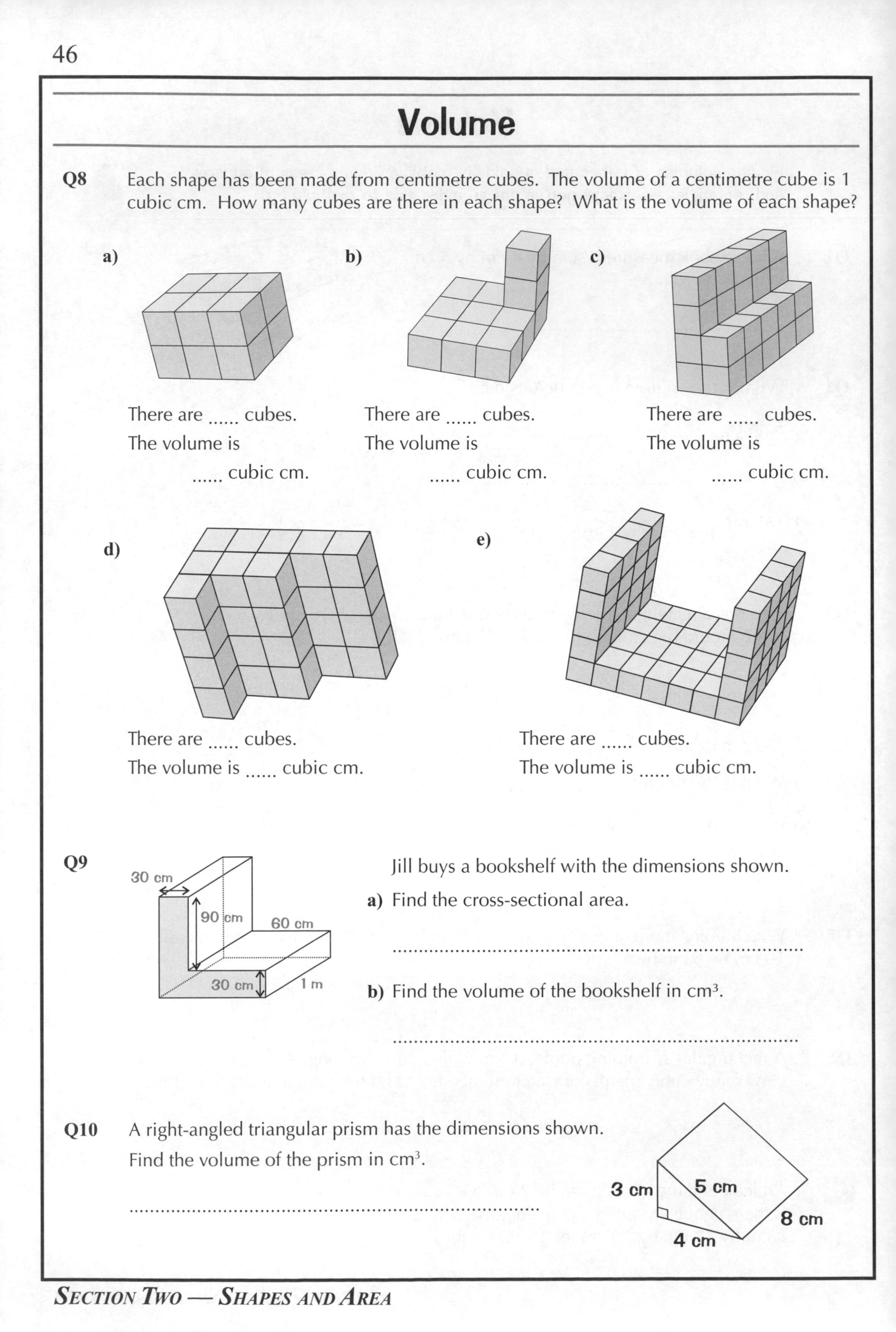

Volume

Q8 Each shape has been made from centimetre cubes. The volume of a centimetre cube is 1 cubic cm. How many cubes are there in each shape? What is the volume of each shape?

a) There are ...... cubes.
The volume is
...... cubic cm.

b) There are ...... cubes.
The volume is
...... cubic cm.

c) There are ...... cubes.
The volume is
...... cubic cm.

d) There are ...... cubes.
The volume is ...... cubic cm.

e) There are ...... cubes.
The volume is ...... cubic cm.

Q9

Jill buys a bookshelf with the dimensions shown.

a) Find the cross-sectional area.

....

b) Find the volume of the bookshelf in cm^3.

....

Q10 A right-angled triangular prism has the dimensions shown.
Find the volume of the prism in cm^3.

....

Volume

Contrary to popular belief, there isn't anything that complicated about prisms — they're only solids with the same shape all the way through. The only bit that sometimes takes a bit longer is finding the cross-sectional area. (Back a few pages for a reminder of areas.)

Volume of any prism = Cross-sectional area × Length

Constant Area of Cross-section

Length

Q11 A tree trunk can be thought of as a circular prism with a height of 1.7 m. If the trunk has a diameter of 60 cm what volume of wood is this in m^3?

....

Q12 A coffee mug is a cylinder closed at one end. The internal radius is 7 cm and the internal height is 9 cm.

a) Taking π to be 3.14, find the volume of liquid the mug can hold.

....

b) If 1200 cm^3 of liquid is poured into the mug, find the depth to the nearest whole cm.

....

....

The depth is just the length of mug taken up by the liquid — which you find by rearranging the volume formula.

Q13 An unsharpened pencil can be thought of as a regular hexagonal prism with a cylinder of graphite through the centre.

You need to use a bit of Pythagoras to find the height of the triangles.

a) By considering a hexagon to be made up of six equilateral triangles, calculate the area of the cross-section of the hexagonal prism shown.

....

circle 2mm diameter

hexagon 4mm each side

b) Find the area of wood in the cross-section.

....

c) If the pencil is 20 cm long what is the volume of wood in the pencil?

....

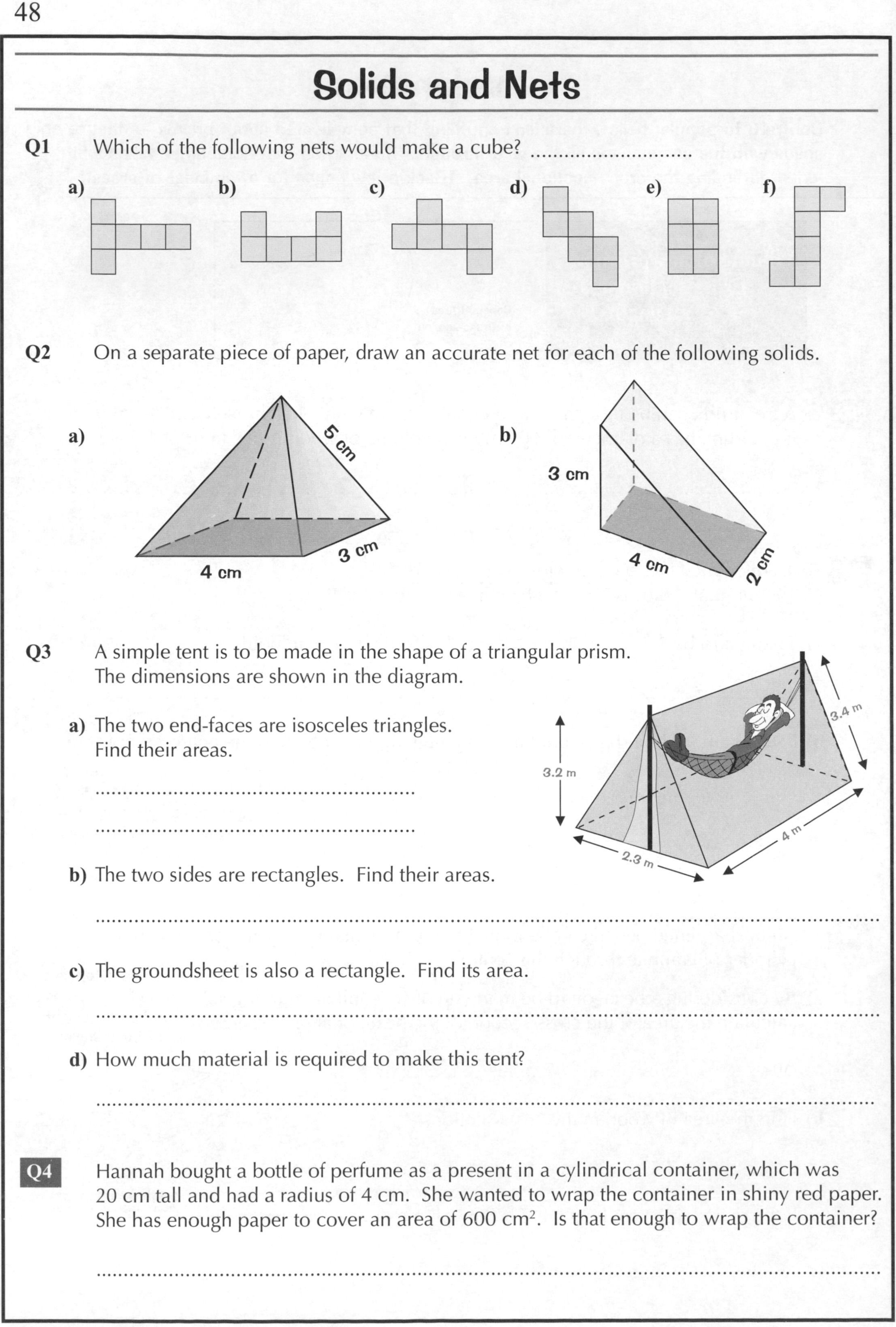

Solids and Nets

Q1 Which of the following nets would make a cube? ....

a) **b)** **c)** **d)** **e)** **f)**

Q2 On a separate piece of paper, draw an accurate net for each of the following solids.

a)

b)

Q3 A simple tent is to be made in the shape of a triangular prism. The dimensions are shown in the diagram.

a) The two end-faces are isosceles triangles. Find their areas.

....

....

b) The two sides are rectangles. Find their areas.

....

c) The groundsheet is also a rectangle. Find its area.

....

d) How much material is required to make this tent?

....

Q4 Hannah bought a bottle of perfume as a present in a cylindrical container, which was 20 cm tall and had a radius of 4 cm. She wanted to wrap the container in shiny red paper. She has enough paper to cover an area of 600 cm^2. Is that enough to wrap the container?

....

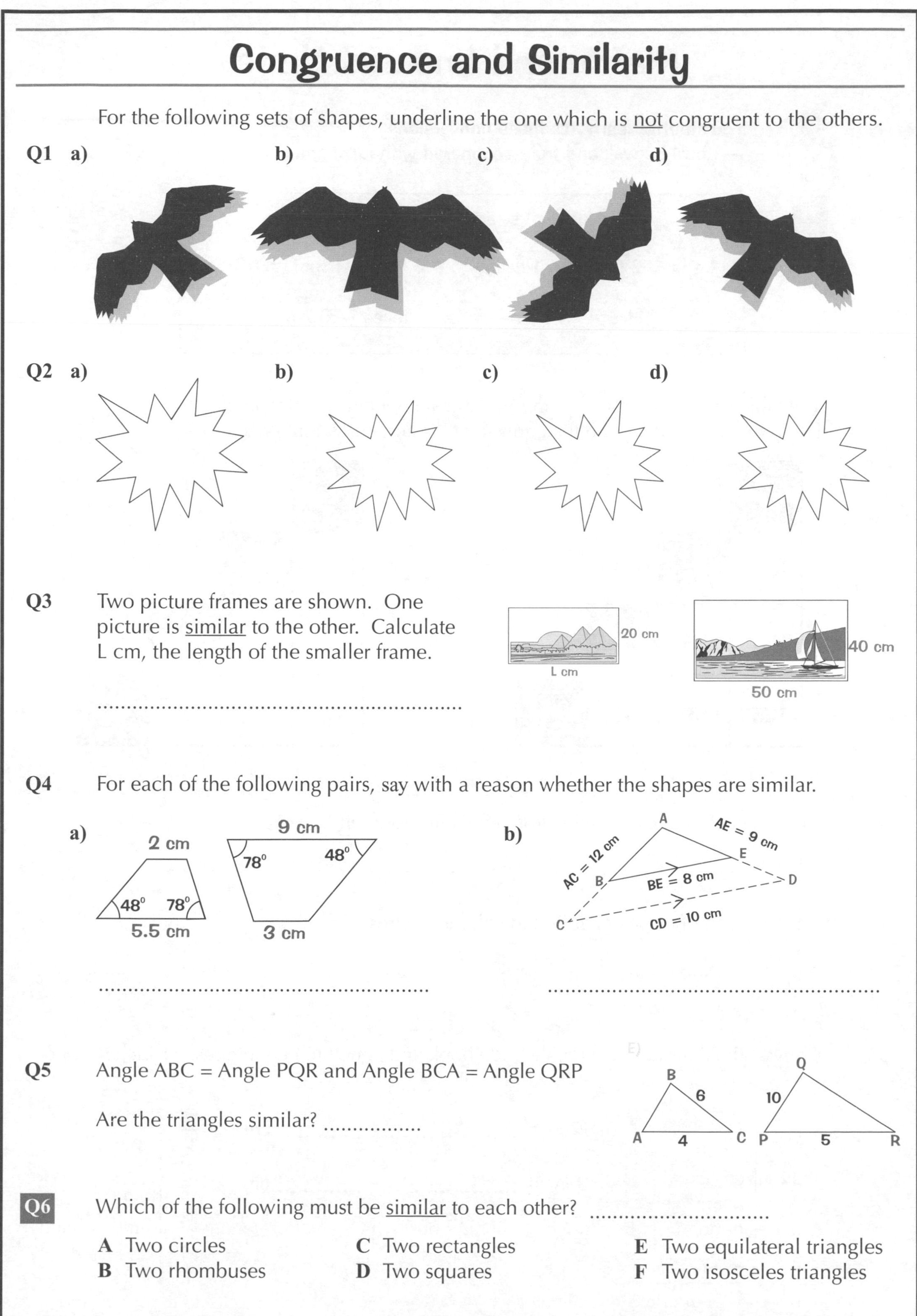

Congruence and Similarity

For the following sets of shapes, underline the one which is not congruent to the others.

Q1 a) b) c) d)

Q2 a) b) c) d)

Q3 Two picture frames are shown. One picture is similar to the other. Calculate L cm, the length of the smaller frame.

....

Q4 For each of the following pairs, say with a reason whether the shapes are similar.

a)

b)

....

Q5 Angle ABC = Angle PQR and Angle BCA = Angle QRP

Are the triangles similar? ....

Q6 Which of the following must be similar to each other? ....

A Two circles
B Two rhombuses
C Two rectangles
D Two squares
E Two equilateral triangles
F Two isosceles triangles

Metric and Imperial Units

You'd better learn ALL these conversions — you'll be well and truly scuppered without them.

APPROXIMATE CONVERSIONS

1 kg = 2.2 lbs 1 gallon = 4.5 l 1 foot = 30 cm

1 litre = 1.75 pints 5 miles = 8 km

Q1 The table shows the distances in miles between 4 cities in Scotland.
Fill in the blank table with the equivalent distances in kilometres.

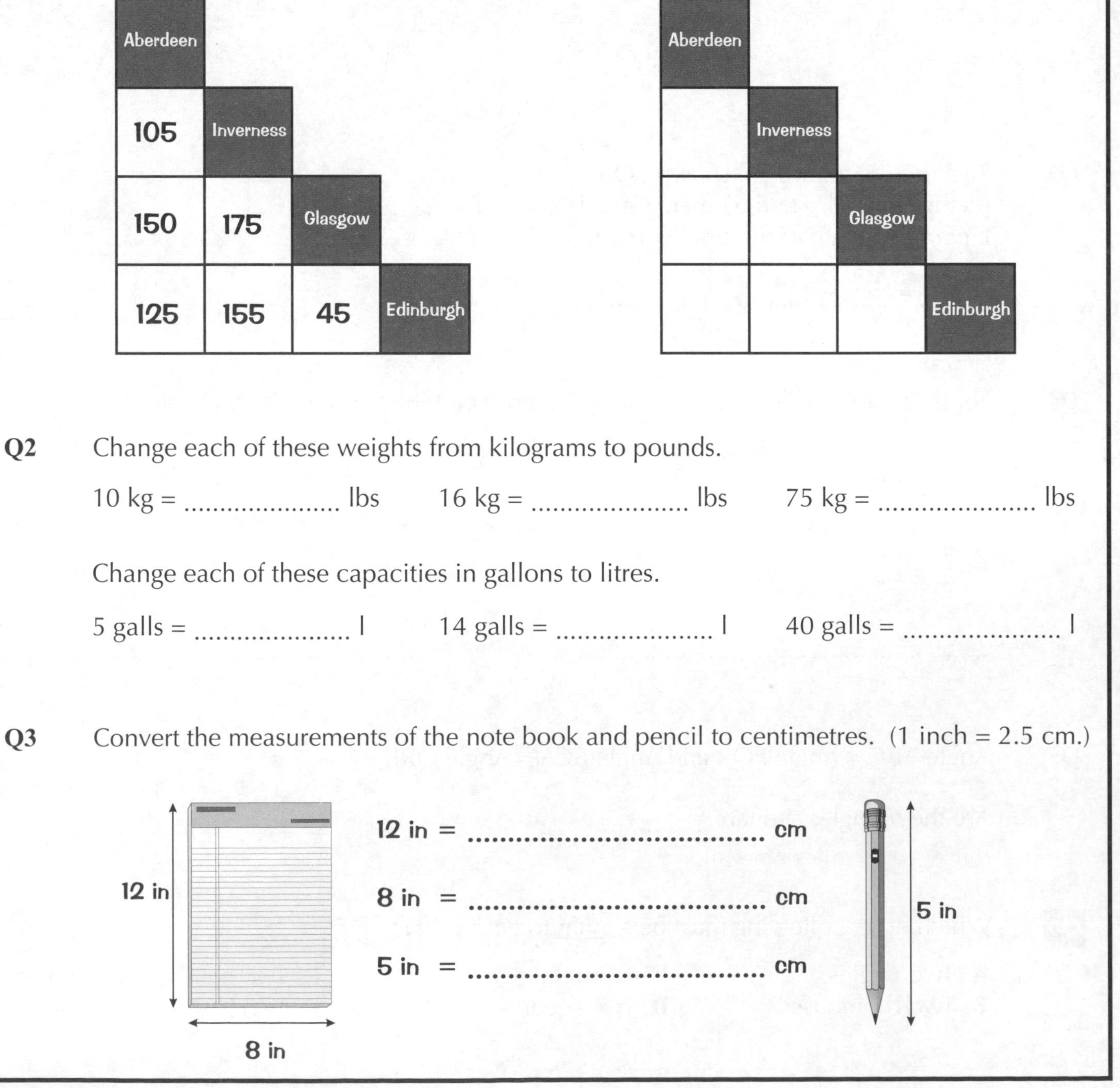

Aberdeen			
105	Inverness		
150	175	Glasgow	
125	155	45	Edinburgh

Aberdeen			
	Inverness		
		Glasgow	
			Edinburgh

Q2 Change each of these weights from kilograms to pounds.

10 kg = ... lbs 16 kg = ... lbs 75 kg = ... lbs

Change each of these capacities in gallons to litres.

5 galls = ... l 14 galls = ... l 40 galls = ... l

Q3 Convert the measurements of the note book and pencil to centimetres. (1 inch = 2.5 cm.)

12 in = ... cm

8 in = ... cm

5 in = ... cm

Metric and Imperial Units

Q4 The water butt in my garden holds 20 gallons of rain-water. How many litres is this?

....

Q5 Tom walked 17 km in one day, while Paul walked 10 miles. Who walked further?

....

....

It doesn't matter which distance you convert — but here it's easier to convert Paul's miles to km.

Q6 A recipe for strawberry jam requires 4 lb of sugar. How many 1 kg bags of sugar does Sarah need to buy so that she can make the jam?

....

Q7 David is throwing a party for himself and 15 of his friends. He decides that it would be nice to make a bowl of fruit punch and carefully follows a recipe for 8 pints worth.

a) Will David fit all the punch in his 5 litre bowl?

....

Read the questions carefully — it's ml for part b), then back to pints for part c).

b) Is there enough punch for everyone at the party to have one 300 ml glass?

....

c) In the end, 12 people split the punch equally between them. How many pints did they each drink?

....

Q8 The average fuel efficiency of two cars is shown below. Which car is the most efficient?
Car A — 53 miles per gallon
Car B — 4.5 litres per 100 km

Remember: 4.5 litres = 1 gallon

....

....

Conversion Factors

The method for these questions is very easy so you might as well learn it...

1) Find the Conversion Factor
2) Multiply by it AND divide by it
3) Choose the common sense answer

The conversion factor is the link between the two units — e.g. there are 100 cm in a m so the conversion factor is 100.

Q1 Fill in the gaps using the conversion factors:

20 mm = ... cm .... mm = 6 cm 3470 m = ... km

.... m = 2 km 3 km = ... cm .... mm = 3.4 m

8550 g = ... kg 1.2 l = ... ml 4400 ml = ... l

Q2 Justin is shopping online.
He looks up the following exchange rates:

1.60 US Dollars ($) to £1 Sterling.
150 Japanese Yen (¥) to £1 Sterling.
2 Australian Dollars (AUD) to £1 Sterling.

Use these exchange rates to calculate to the nearest penny the cost in Sterling of each of Justin's purchases:

a) A book costing $7.50

....

b) An MP3 player costing ¥7660

....

c) An electric guitar costing 683 AUD

....

d) Justin has two quotes for the cost of shipping his guitar from Australia to the UK:
155 AUD from an Australian courier and £76.45 from a British courier.
Which company is cheaper?

....

Q3 Jeremy used a website to calculate his car's efficiency. He needed to enter the engine capacity in cubic centimetres, but only knew that it was 1.4 litres.

What capacity should he have entered into the website?

Remember, 1 ml = 1 cm^3

....

Q4 Chris wants to varnish the surface of his table, which has an area of 1200 cm^2.
He buys a tin of varnish that says it will cover an area of 2 m^2.

Find the area of the table surface in m^2.
Does Chis have enough varnish to cover the surface of his table?

....

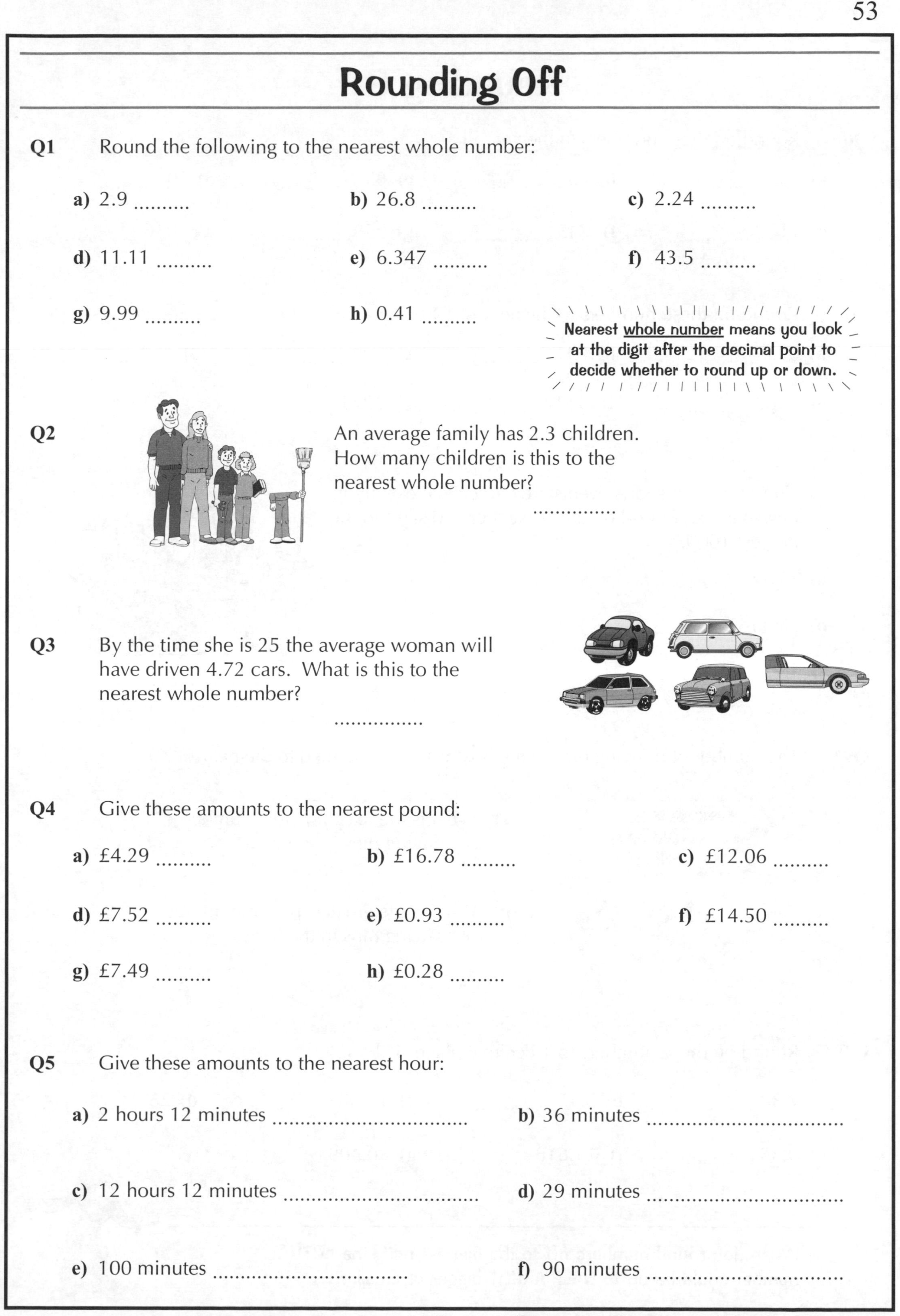

Rounding Off

Q1 Round the following to the nearest whole number:

a) 2.9 ....
b) 26.8 ....
c) 2.24 ....
d) 11.11 ....
e) 6.347 ....
f) 43.5 ....
g) 9.99 ....
h) 0.41 ....

Nearest whole number means you look at the digit after the decimal point to decide whether to round up or down.

Q2 An average family has 2.3 children. How many children is this to the nearest whole number?

....

Q3 By the time she is 25 the average woman will have driven 4.72 cars. What is this to the nearest whole number?

....

Q4 Give these amounts to the nearest pound:

a) £4.29 ....
b) £16.78 ....
c) £12.06 ....
d) £7.52 ....
e) £0.93 ....
f) £14.50 ....
g) £7.49 ....
h) £0.28 ....

Q5 Give these amounts to the nearest hour:

a) 2 hours 12 minutes ....
b) 36 minutes ....
c) 12 hours 12 minutes ....
d) 29 minutes ....
e) 100 minutes ....
f) 90 minutes ....

Rounding Off

Q6 Round off these numbers to the nearest 10:

a) 23 ....
b) 78 ....
c) 65 ....
d) 99 ....

e) 118 ....
f) 243 ....
g) 958 ....
h) 1056 ....

Q7 Round off these numbers to the nearest 100:

a) 627 ....
b) 791 ....
c) 199 ....
d) 450 ....

e) 1288 ....
f) 3329 ....
g) 2993 ....

Q8 Crowd sizes at sports events are often given exactly in newspapers. Round off these exact crowd sizes to the nearest 1000:

a) 23 324 ....

b) 36 844 ....

c) 49 752 ....

Q9 The number of drawing pins in the box has been rounded to the nearest 10.

DRAWING PINS
Contents: 80

a) What is the least possible number of drawing pins in the box?

....

b) What is the greatest possible number of drawing pins in the box?

....

Q10 Round off these numbers to 1 decimal place (1 d.p.):

a) 7.34 ....
b) 8.47 ....
c) 12.08 ....
d) 28.03 ....

e) 9.35 ....
f) 14.618 ....
g) 30.409 ....

When you round numbers off to the nearest unit, the ACTUAL number could be up to HALF A UNIT bigger or smaller...

Rounding Off

Q11 Round off the following to 2 d.p.:

a) 17.363 .... **b)** 38.057 .... **c)** 0.735 ....

d) 5.99823 .... **e)** 4.297 .... **f)** 7.0409 ....

Q12 Now round these to 3 d.p.:

a) 6.3534 .... **b)** 81.64471 .... **c)** 0.0075 ....

d) 53.26981 .... **e)** 754.39962 .... **f)** 0.000486 ....

Q13 Seven people have a meal in a restaurant. The total bill comes to £60. If they share the bill equally, how much should each of them pay?
Round your answer to 2 d.p.

....

Q14 Round these numbers to 1 significant figure.

a) 12 .... **b)** 530 ....

c) 1379 .... **d)** 0.021 ....

e) 1829.62 .... **f)** 0.296 ....

The 1st significant figure of any number is the first digit which isn't zero.

Q15 Round these numbers to 3 significant figures.

a) 1379 .... **b)** 1329.62 ....

c) 0.296 .... **d)** 0.02139 ....

Q16 $K = 56.762$
Write K correct to:

a) three significant figures ....

b) two significant figures ....

c) one significant figure. ....

Q17 Matt is buying skirting board to fit to a wall which measures 540 cm to 2 s.f. What length of board should Matt buy to guarantee he has enough for the whole wall?

....

Accuracy and Estimating

Round off to NICE EASY CONVENIENT NUMBERS, then use them to do the sum. Easy peasy.

Q1 Estimate the answers to these questions...

For example: 12 × 21 <u>10 × 20</u> = <u>200</u>

a) 18 × 12 .... × .... = ...

b) 23 × 21 .... × .... = ...

c) 57 × 46 .... × .... = ...

d) 98 × 145 .... × .... = ...

e) 11 ÷ 4 .... ÷ .... = ...

f) 22 ÷ 6 .... ÷ .... = ...

g) 97 ÷ 9 .... ÷ .... = ...

h) 147 ÷ 14 .... ÷ .... = ...

i) 195 × 205 .... × .... = ...

j) 545 × 301 .... × .... = ...

k) 901 ÷ 33 .... ÷ .... = ...

l) 1207 ÷ 598 .... ÷ .... = ...

Q2 Andy earns a salary of £24 108 each year.

a) Estimate how much Andy earns per month. £ ....

b) Andy is hoping to get a bonus this year. The bonus is calculated as £1017 plus 9.7% of each worker's salary. Estimate the amount of Andy's bonus.

£ ....

c) Andy pays 10.14% of his regular salary into a pension scheme. Estimate how much money he has left per year after making his pension payments.

£ ....

Q3 Estimate the following lengths then measure them to see how far out you were:

OBJECT	ESTIMATE	ACTUAL LENGTH
a) Length of your pen or pencil	....	....
b) Width of your thumbnail	....	....
c) Height of this page	....	....
d) Length of the room you are in	....	....

Accuracy and Estimating

Q4 The spade is almost 1 m tall.
Estimate the height of the sunflower in metres.

....

If you have trouble estimating the height by eye, try measuring the spade against your finger. Then see how many times that bit of finger fits into the height of the sunflower.

Q5 The distance from Galway to Tuam is approximately 20 miles.
Estimate the distance from Galway to Clifden.

Clifden
Tuam
Lough Corrib
Galway
Galway Bay
Aran Islands

.... miles

Q6 A village green is roughly rectangular with a length of 33 m 48 cm and is 24 m 13 cm wide. Calculate the area of the green in m^2 to:

a) 2 decimal places ....

b) 3 significant figures ....

c) State which of parts **a)** and **b)** would be the more reasonable value to use. ....

Q7 Decide on an appropriate degree of accuracy for the following:

a) the total dry weight, 80 872 kg, of the space shuttle OV-102 Columbia with its 3 main engines ....

b) the distance of 3.872 miles from Mel's house to Bryan's house ....

c) 1.563 m of fabric required to make a bedroom curtain ....

Q8 Calculate, giving your answer to an appropriate degree of accuracy:

a) $\dfrac{41.75 \times 0.9784}{22.3 \times 2.54}$ = ...

b) $\dfrac{12.54 + 7.33}{12.54 - 7.22}$ = ...

Clock Time Questions

I'm sure you know the difference between 12 and 24 hour clocks, but just so there's no excuses...

Q1 The times below are given using the 24 hour system. Using am or pm, give the equivalent time for a 12 hour clock.

a) 0400 .... **b)** 0215 .... **c)** 2130 ....

Q2 The times below are taken from a 12 hour clock. Give the equivalent 24 hour readings.

a) 11.22 am .... **b)** 12.30 pm .... **c)** 3.33 pm ....

Q3 Convert the following into hours and minutes:

a) 3.75 hours .... **b)** 0.2 hours .... **c)** 5.8 hours ....

Q4 Fran is going to cook a 5.5 kg ham. The ham needs to be cooked for 40 minutes per kg, and then left to "rest" for 25 minutes before serving. Fran wants to serve the ham at 4 pm. What time should she start cooking?

....

Q5 Steve sets off on a bike ride at 10.30 am. He stops for lunch at 12.15 pm and sets off again at 1 pm. He has a 20 minute rest stop in the afternoon and gets home at 4.50 pm.

How long did he cycle for altogether? ....

Q6 This timetable refers to three trains that travel from Asham to Derton.

a) Which train is quickest from Asham to Derton? ....

b) Which train is quickest from Cottingham to Derton?

....

c) I live in Bordhouse. It takes me 8 minutes to walk to the train station. At what time must I leave the house by to arrive in Derton before 2.30 pm?

Asham – Derton	Train 1	Train 2	Train 3
Asham	0832	1135	1336
Bordhouse	0914	1216	1414
Cottingham	1002	1259	1456
Derton	1101	1404	1602

....

Compass Directions and Bearings

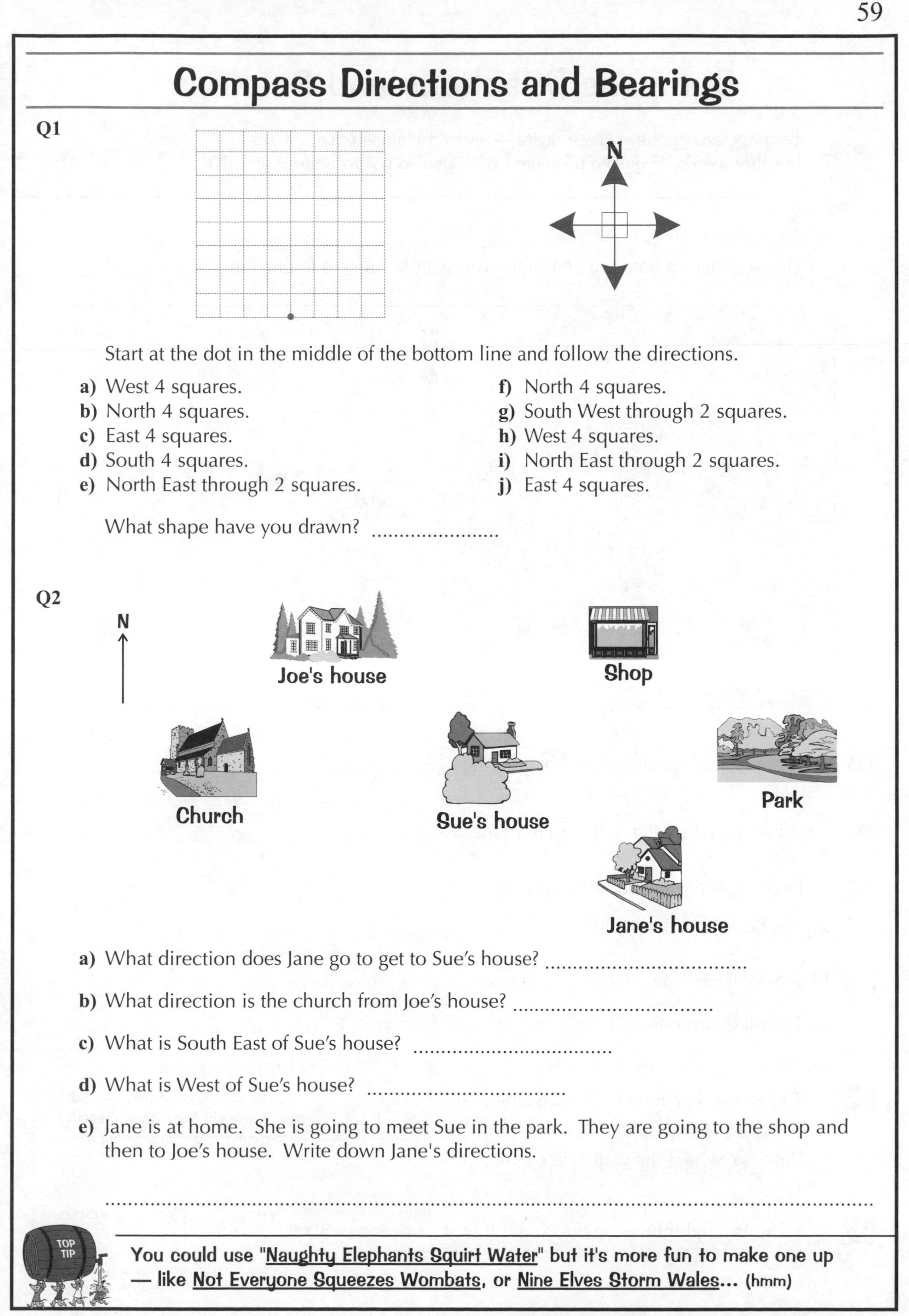

Q1

N

Start at the dot in the middle of the bottom line and follow the directions.

a) West 4 squares.
b) North 4 squares.
c) East 4 squares.
d) South 4 squares.
e) North East through 2 squares.
f) North 4 squares.
g) South West through 2 squares.
h) West 4 squares.
i) North East through 2 squares.
j) East 4 squares.

What shape have you drawn? ....

Q2

N

Joe's house

Shop

Church

Sue's house

Park

Jane's house

a) What direction does Jane go to get to Sue's house? ....

b) What direction is the church from Joe's house? ....

c) What is South East of Sue's house? ....

d) What is West of Sue's house? ....

e) Jane is at home. She is going to meet Sue in the park. They are going to the shop and then to Joe's house. Write down Jane's directions.

....

TOP TIP

You could use "Naughty Elephants Squirt Water" but it's more fun to make one up — like Not Everyone Squeezes Wombats, or Nine Elves Storm Wales... (hmm)

Compass Directions and Bearings

**Bearings always have three digits — even the small ones...
in other words, if you've measured 60°, you've got to write it as 060°.**

This is a map of part of a coastline. The scale is one cm to one km.

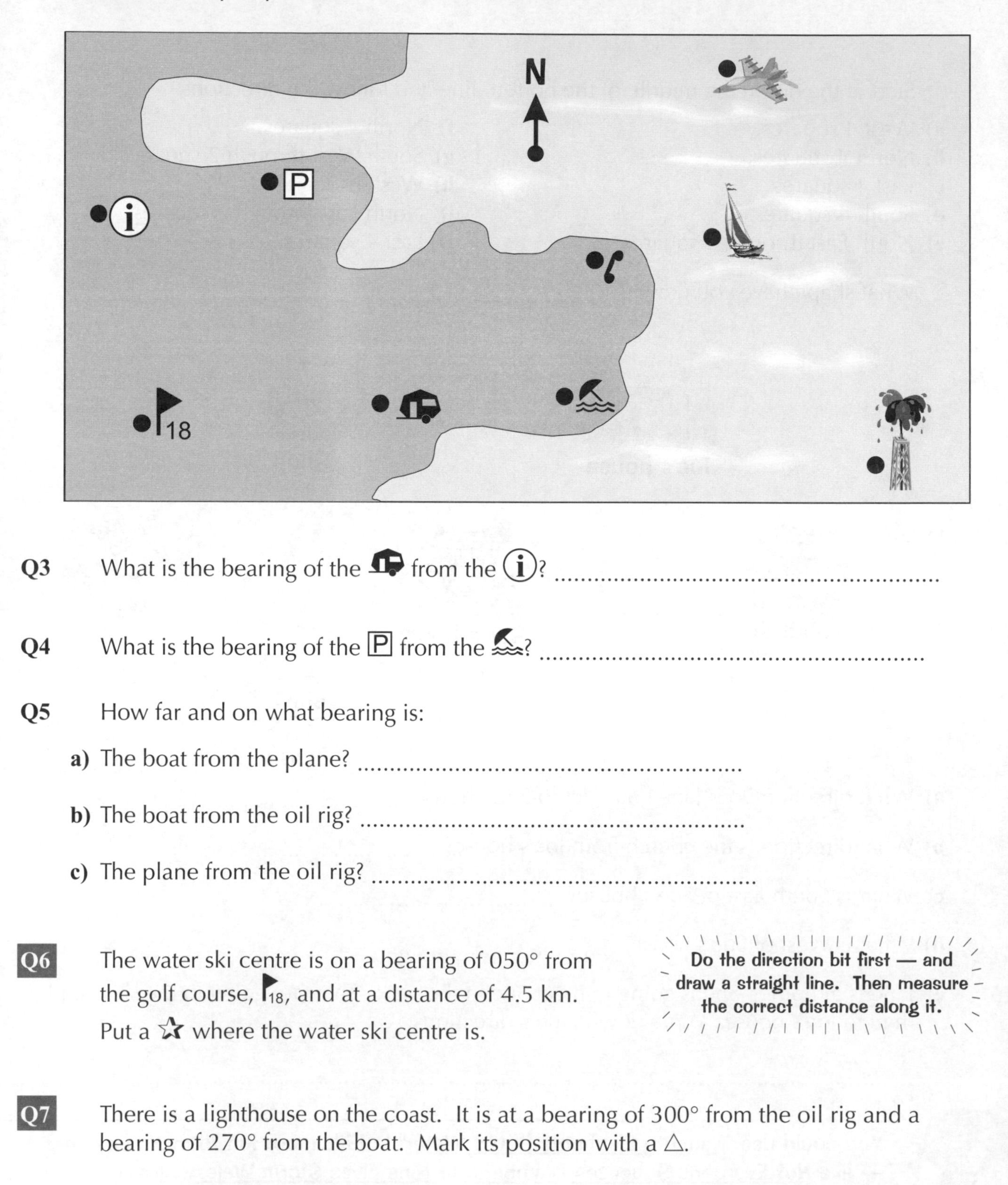

Q3 What is the bearing of the from the i? ....

Q4 What is the bearing of the P from the ? ....

Q5 How far and on what bearing is:

a) The boat from the plane? ....

b) The boat from the oil rig? ....

c) The plane from the oil rig? ....

Q6 The water ski centre is on a bearing of 050° from the golf course, 18, and at a distance of 4.5 km. Put a ☆ where the water ski centre is.

Do the direction bit first — and draw a straight line. Then measure the correct distance along it.

Q7 There is a lighthouse on the coast. It is at a bearing of 300° from the oil rig and a bearing of 270° from the boat. Mark its position with a △.

Maps and Map Scales

Q1 The scale on this map is 1 cm : 4 km.

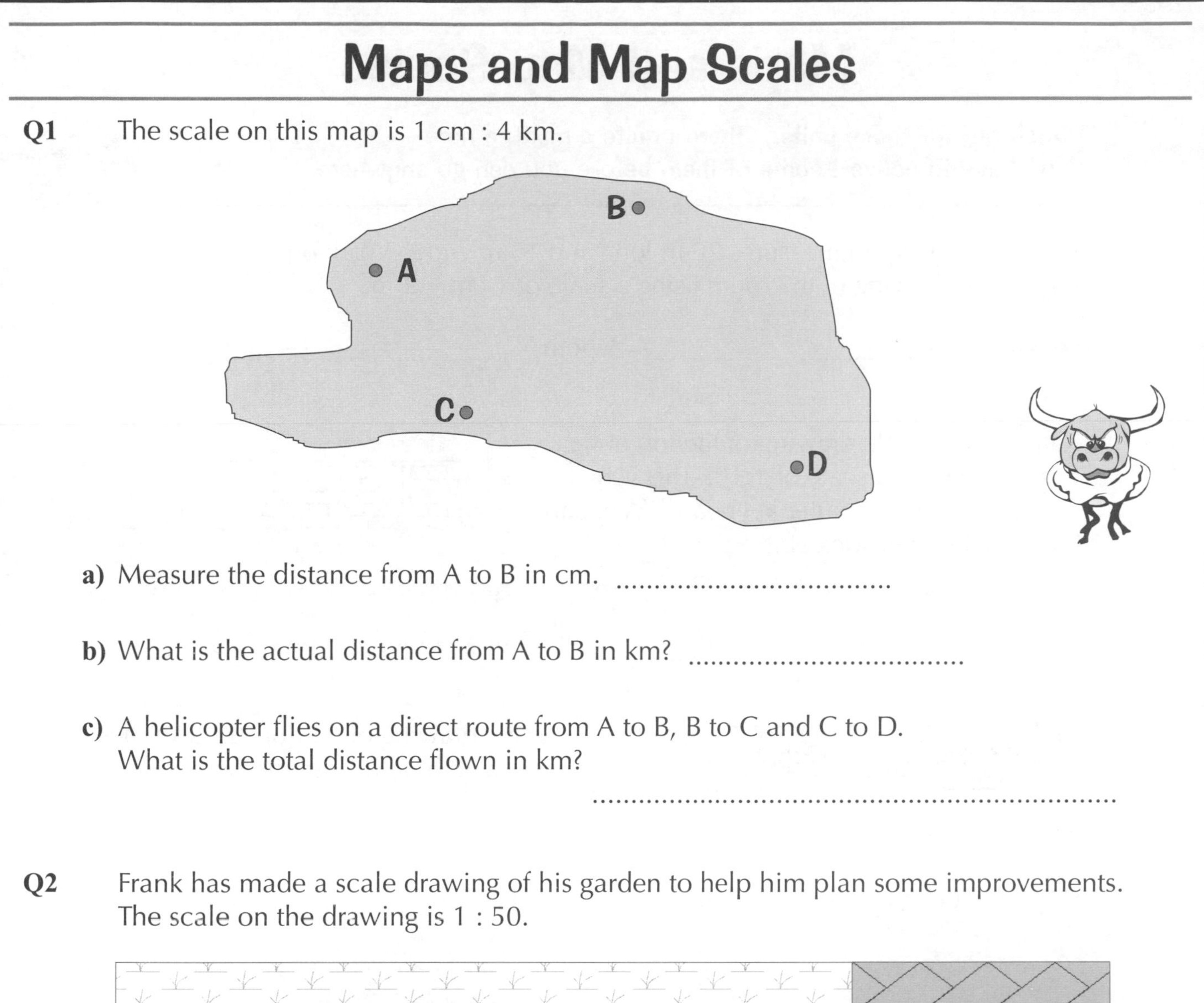

a) Measure the distance from A to B in cm. ....

b) What is the actual distance from A to B in km? ....

c) A helicopter flies on a direct route from A to B, B to C and C to D.
What is the total distance flown in km?

....

Q2 Frank has made a scale drawing of his garden to help him plan some improvements.
The scale on the drawing is 1 : 50.

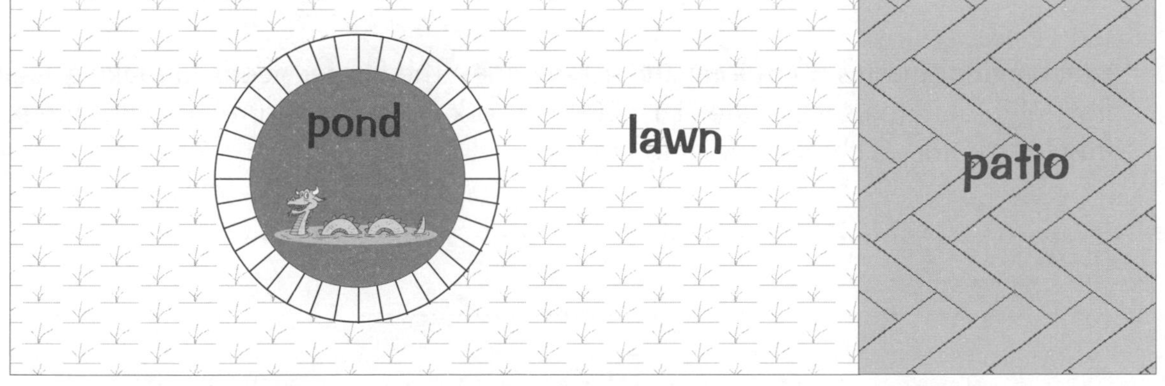

a) Frank wants to put up a fence along the three outside edges of the lawn.
How many metres of fencing does he need to buy?

....

b) What are the actual dimensions of Frank's patio in m?

....

If the scale doesn't say what units it's in, it just means that both sides of the ratio are the same units — so <u>1 : 1000</u> would, for example, mean <u>1 cm : 1000 cm</u>.

Maps and Map Scales

Watch out for those units... there's quite a mixture here — you'll have to convert some of them before you can go anywhere.

Q3 A rectangular room measures 20 m long and 15 m wide. Work out the measurements for a scale drawing of the room using a scale of 1 cm = 2 m.

Length = ... Width = ...

Q4 Katie drew a scale drawing of the top of her desk. She used a scale of 1:10. This is her drawing of the computer keyboard. What are the actual dimensions of it?

Length = ... Width = ...

Q5 This is a scale drawing of part of Paul's kitchen. Measure the width of the gap for the oven.

Cupboards | Oven | Cupboards

.... mm.

The drawing uses a scale of 1 : 60.
Work out the maximum width of oven, in cm, that Paul can buy.

....

Q6 A rectangular field is 60 m long and 40 m wide. The farmer needs to make a scale drawing of it. He uses a scale of 1 : 2000. Work out the measurements for the scale drawing. (Hint — change the m to cm.)

....

....

Q7 A rectangular room is 4.8 m long and 3.6 m wide. Make a scale drawing of it using a scale of 1 cm to 120 cm. First work out the measurements for the scale drawing.

Length = ...

Width = ...

On your scale drawing mark a window, whose actual length is 2.4 m, on one long wall and mark a door, actual width 90 cm, on one short wall.

Window = ...

Door = ...

Maps and Directions

Remember this great way of remembering what order to read grid references: First Across the bottom, Then Up the side. FAT Uncle (or FAT Unicorn, if you like).

Q1 Below is a map of Damcoster town centre.

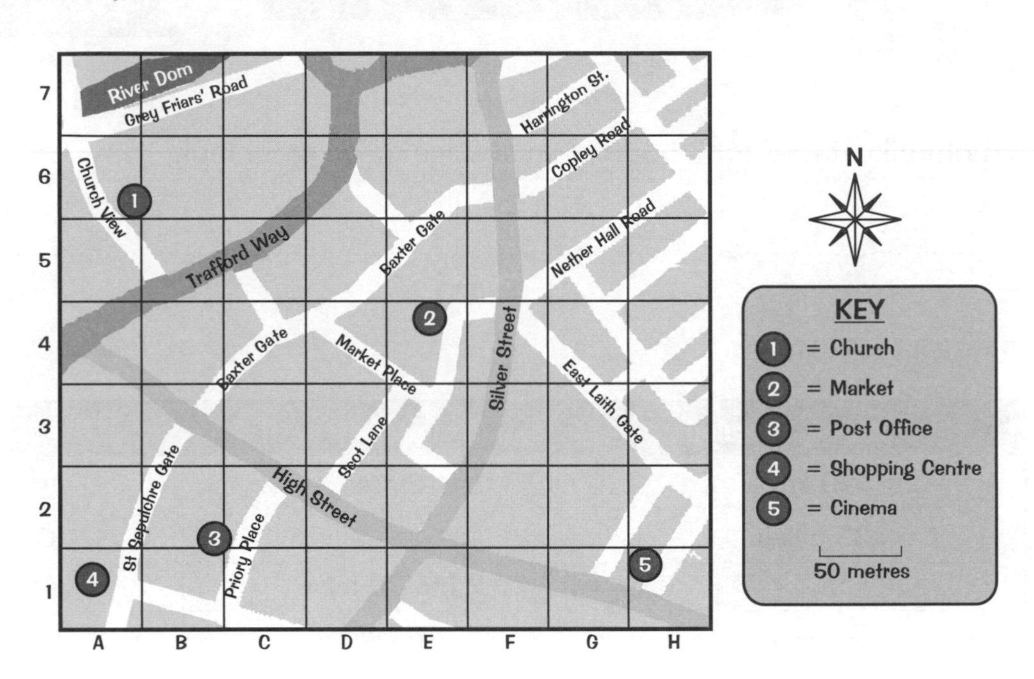

a) What compass direction is the market from the shopping centre?

....

b) Ric walks East from the market. When he reaches Silver Street, he walks North East up the street and takes his first left. He then takes his second left.
What street is he now standing upon?

....

c) A group of tourists at the post office want to know where they can get a cappuccino and blueberry muffin. Give them directions to the coffee morning at the local church.

....

....

....

....

d) What is the grid reference of the cinema? ....

e) Estimate the distance to walk from the post office to the cinema. ....

Speed

This is an easy enough formula — and of course you can put it in that good old formula triangle as well.

Average speed = $\frac{\text{Total distance}}{\text{Total time}}$

d
s × t

Q1 A motorbike travels for 3 hours at an average speed of 55 mph. How far has it travelled?

....

Q2 Complete this table:

Distance Travelled	Time taken	Average Speed
210 km	3 hrs	
135 miles		30 mph
	2 hrs 30 mins	42 km/h
9 miles	45 mins	
640 km		800 km/h
	1 hr 10 mins	60 mph

Q3 An athlete can run 100 m in 11 seconds.
Calculate the athlete's speed in:

a) m/s

....

b) km/h

....

Q4 Simon is driving on the motorway at an average speed of 67 mph.
He sees a sign telling him that he is 36 miles away from the next service station.

a) To the nearest minute, how long will it take Simon to reach the service station?

....

b) Later, Simon drove through some roadworks where the speed limit was 50 mph.
Two cameras recorded the time taken to travel 1200 metres through the roadworks as 56 seconds. Was Simon speeding through the roadworks?

....

Remember — 1 mile = 1.6 km

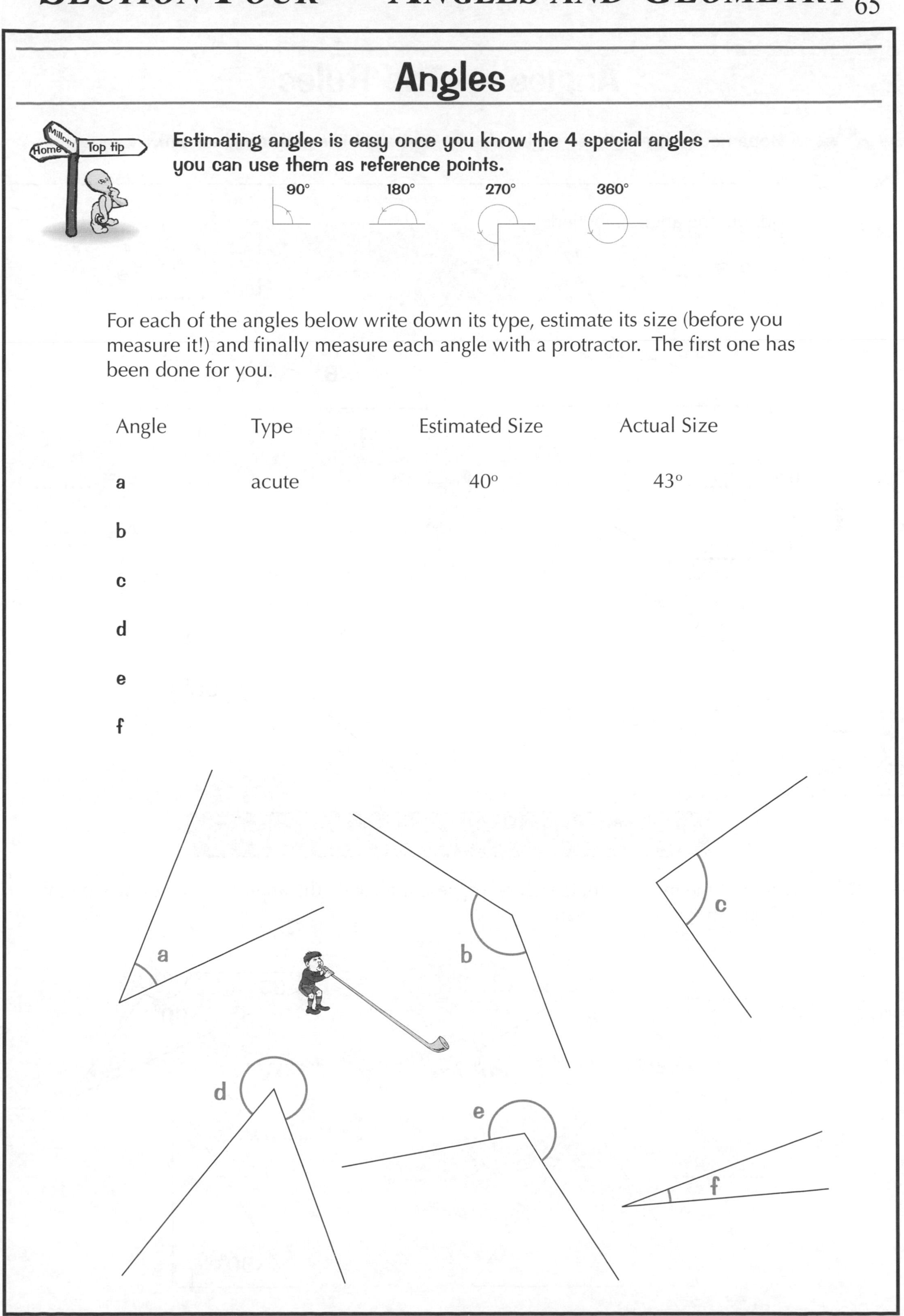

Angles

Estimating angles is easy once you know the 4 special angles — you can use them as reference points.

For each of the angles below write down its type, estimate its size (before you measure it!) and finally measure each angle with a protractor. The first one has been done for you.

Angle	Type	Estimated Size	Actual Size
a	acute	40°	43°
b			
c			
d			
e			
f			

Angles — The Rules

Hope you've learnt those angle rules for a straight line and round a point...

Q1 Work out the angles labelled:

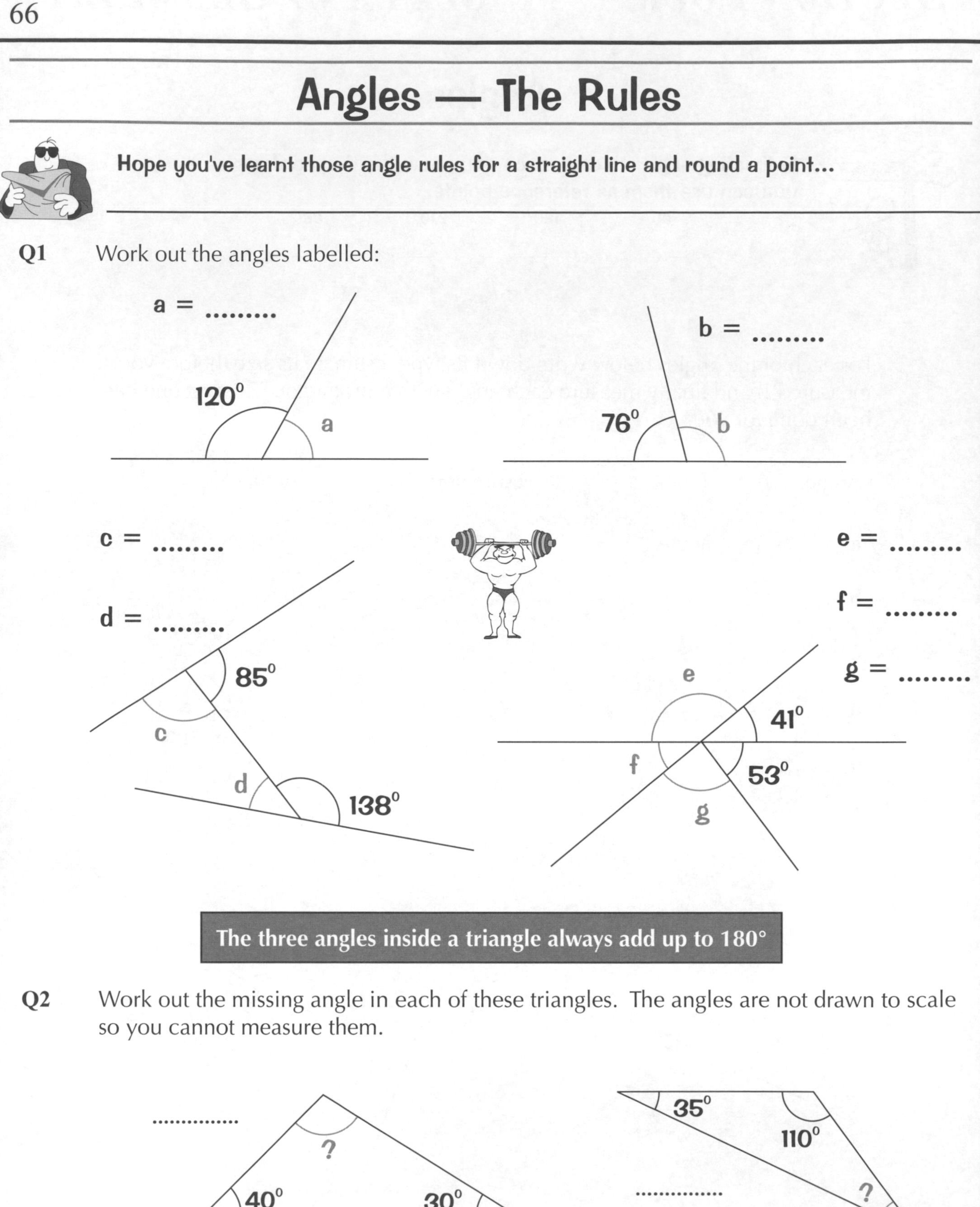

The three angles inside a triangle always add up to 180°

Q2 Work out the missing angle in each of these triangles. The angles are not drawn to scale so you cannot measure them.

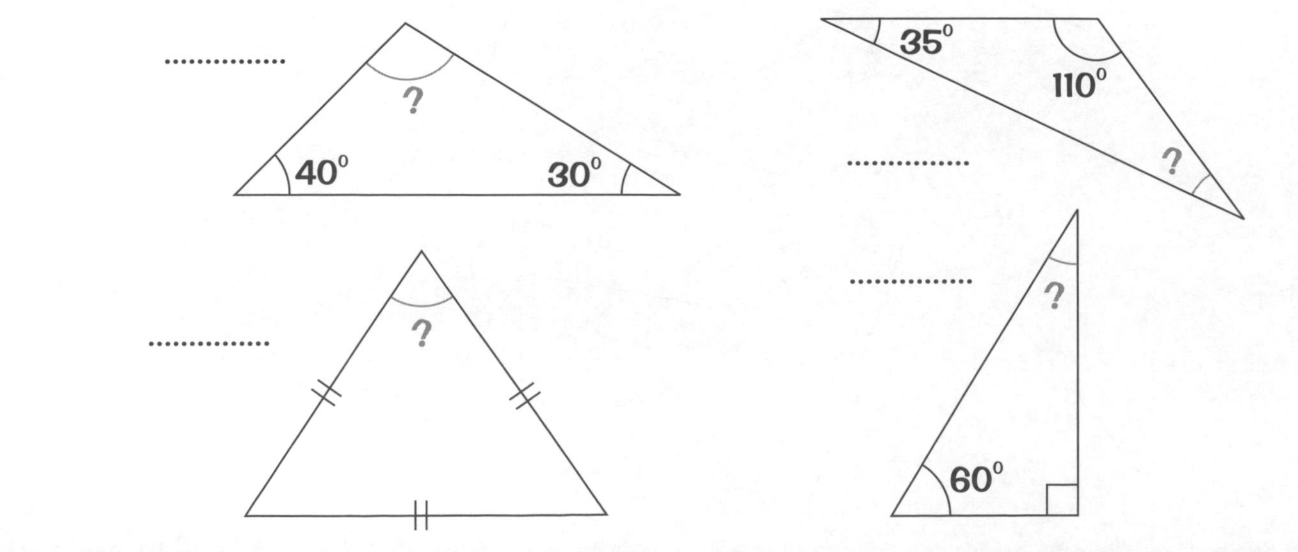

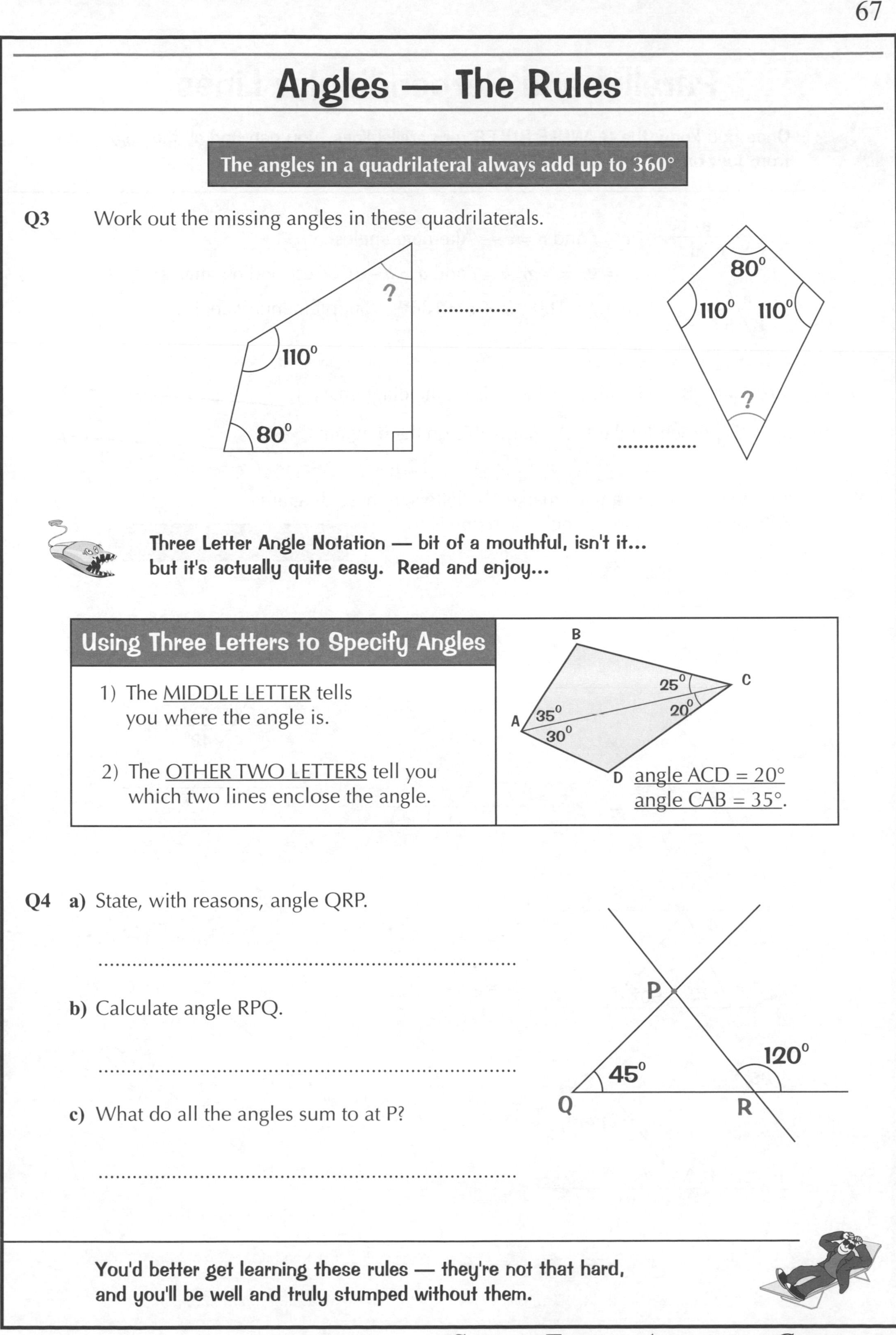

Angles — The Rules

The angles in a quadrilateral always add up to 360°

Q3 Work out the missing angles in these quadrilaterals.

Three Letter Angle Notation — bit of a mouthful, isn't it... but it's actually quite easy. Read and enjoy...

Using Three Letters to Specify Angles

1) The <u>MIDDLE LETTER</u> tells you where the angle is.

2) The <u>OTHER TWO LETTERS</u> tell you which two lines enclose the angle.

<u>angle ACD = 20°</u>
<u>angle CAB = 35°</u>.

Q4 a) State, with reasons, angle QRP.

....

b) Calculate angle RPQ.

....

c) What do all the angles sum to at P?

....

You'd better get learning these rules — they're not that hard, and you'll be well and truly stumped without them.

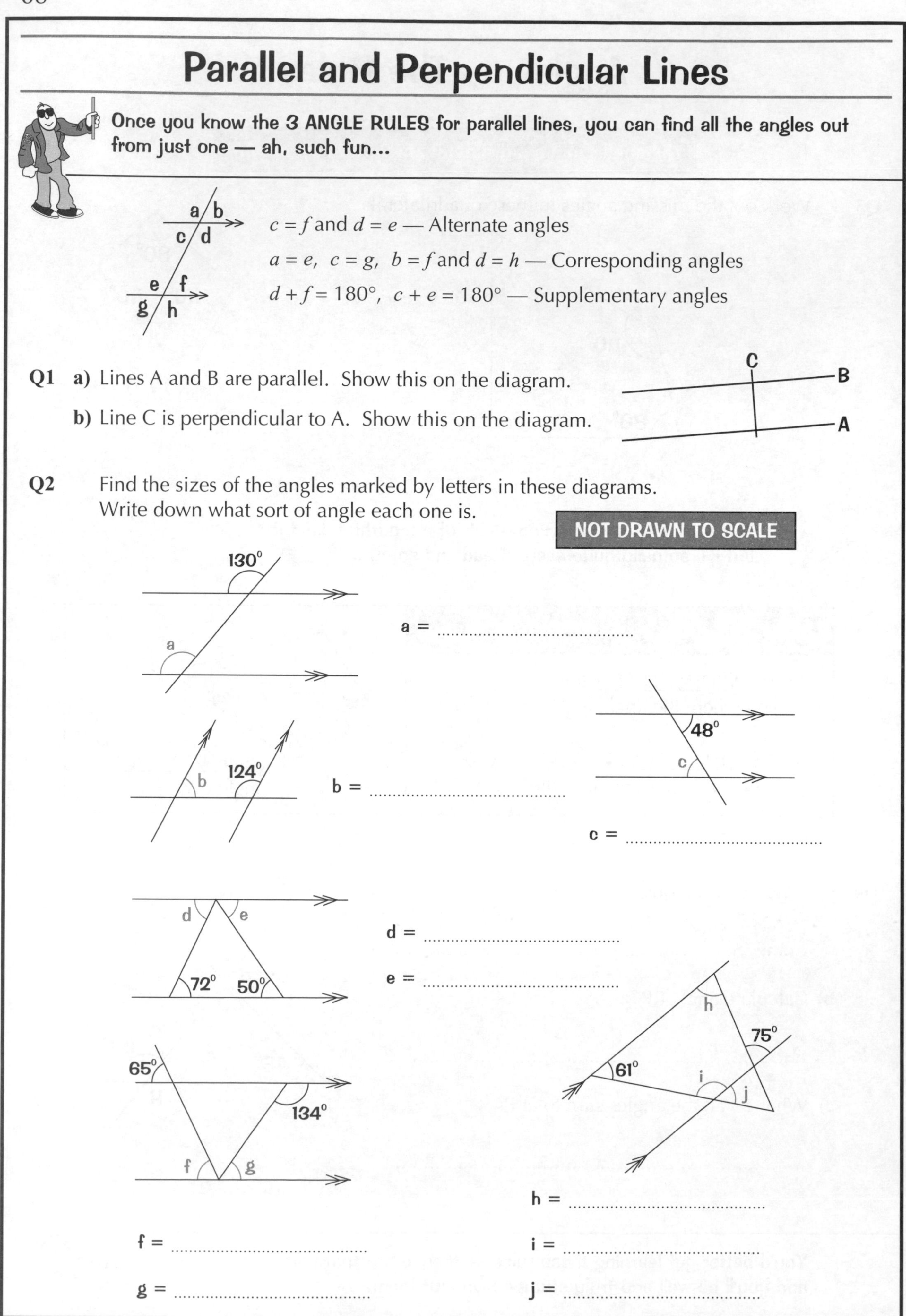

Parallel and Perpendicular Lines

Once you know the 3 ANGLE RULES for parallel lines, you can find all the angles out from just one — ah, such fun...

$c = f$ and $d = e$ — Alternate angles

$a = e$, $c = g$, $b = f$ and $d = h$ — Corresponding angles

$d + f = 180°$, $c + e = 180°$ — Supplementary angles

Q1 **a)** Lines A and B are parallel. Show this on the diagram.

b) Line C is perpendicular to A. Show this on the diagram.

Q2 Find the sizes of the angles marked by letters in these diagrams. Write down what sort of angle each one is.

NOT DRAWN TO SCALE

a = ...

b = ...

c = ...

d = ...

e = ...

h = ...

i = ...

j = ...

f = ...

g = ...

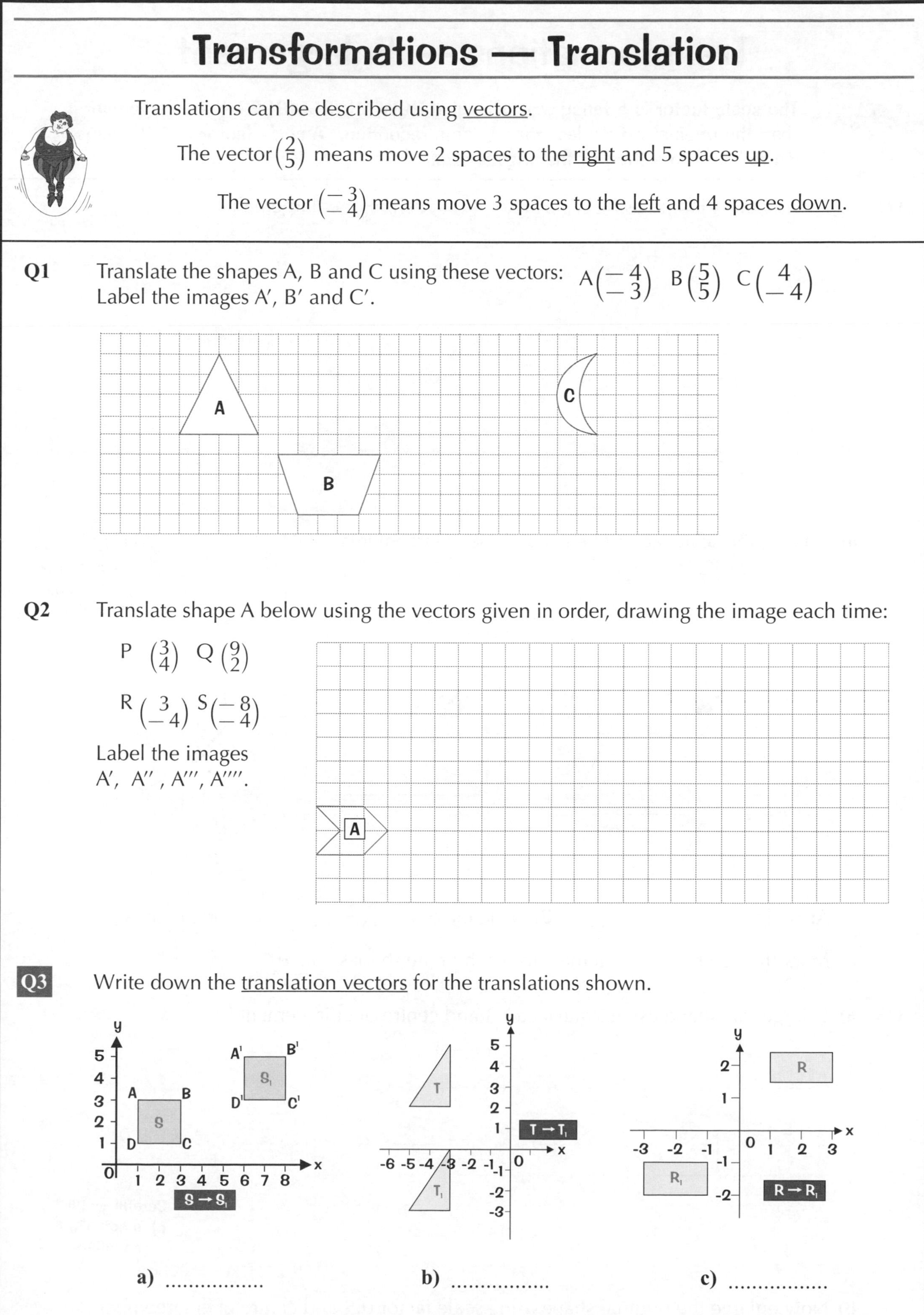

Transformations — Translation

Translations can be described using <u>vectors</u>.

The vector $\begin{pmatrix}2\\5\end{pmatrix}$ means move 2 spaces to the <u>right</u> and 5 spaces <u>up</u>.

The vector $\begin{pmatrix}-3\\-4\end{pmatrix}$ means move 3 spaces to the <u>left</u> and 4 spaces <u>down</u>.

Q1 Translate the shapes A, B and C using these vectors: A$\begin{pmatrix}-4\\-3\end{pmatrix}$ B$\begin{pmatrix}5\\5\end{pmatrix}$ C$\begin{pmatrix}4\\-4\end{pmatrix}$
Label the images A′, B′ and C′.

Q2 Translate shape A below using the vectors given in order, drawing the image each time:

P $\begin{pmatrix}3\\4\end{pmatrix}$ Q $\begin{pmatrix}9\\2\end{pmatrix}$

R $\begin{pmatrix}3\\-4\end{pmatrix}$ S $\begin{pmatrix}-8\\-4\end{pmatrix}$

Label the images A′, A″, A‴, A⁗.

Q3 Write down the <u>translation vectors</u> for the translations shown.

a) .... **b)** .... **c)** ....

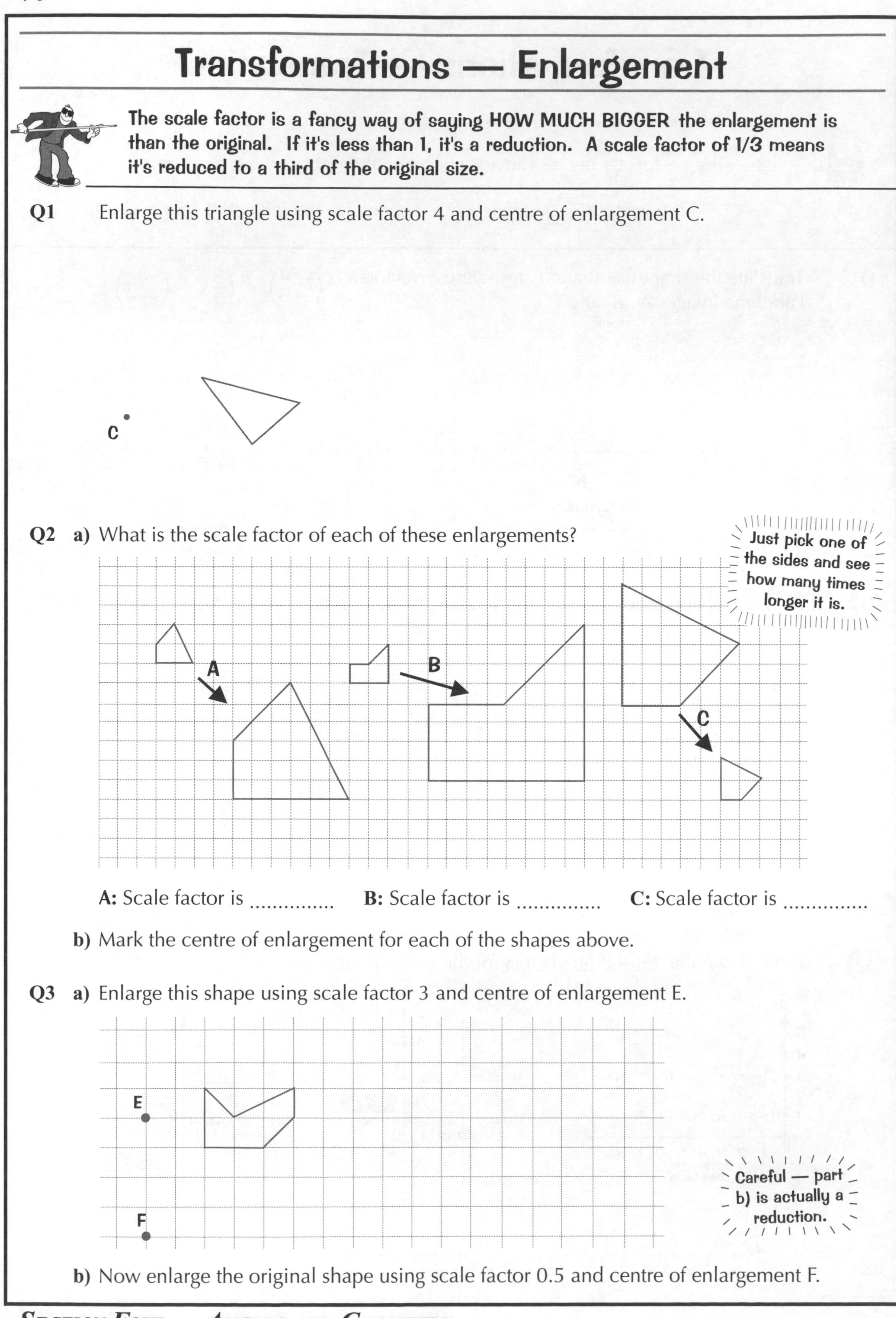

Transformations — Enlargement

The scale factor is a fancy way of saying HOW MUCH BIGGER the enlargement is than the original. If it's less than 1, it's a reduction. A scale factor of 1/3 means it's reduced to a third of the original size.

Q1 Enlarge this triangle using scale factor 4 and centre of enlargement C.

Q2 a) What is the scale factor of each of these enlargements?

A: Scale factor is .... **B:** Scale factor is .... **C:** Scale factor is ....

b) Mark the centre of enlargement for each of the shapes above.

Q3 a) Enlarge this shape using scale factor 3 and centre of enlargement E.

b) Now enlarge the original shape using scale factor 0.5 and centre of enlargement F.

Transformations — Enlargement

Q4 The side view of a playground swing is shown in the diagram. Triangle PST is an enlargement of triangle PQR.

P, 1.5 m, 1.5 m, Q, R, 0.9 m, 3 m, S, T

a) Write down the distance PT.

....

b) Calculate the distance ST.

....

Q5 Jenny has a photo on her computer which has a width of 640 pixels and a height of 480 pixels. She wants to enlarge the photo to use as her computer desktop background. The image must fill her monitor screen, which has a width of 1024 pixels.

a) Calculate the scale factor Jenny must use to enlarge her photo.

....

....

b) Work out the height of the enlarged photo, in pixels.

....

Q6 The large rectangles in diagrams A and B (below) are similar. The small rectangles in the diagrams below are also similar. Find the shaded area of B.

A: 2 m, 3 m, 5 m, 6 m

B: 6 m, 10 m

....

Q7 Two cups A and B are similar. Cup A has a height of 15 cm and cup B has a height of 5 cm. Cup A has a volume of 54 cm^3. Calculate the volume of cup B.

....

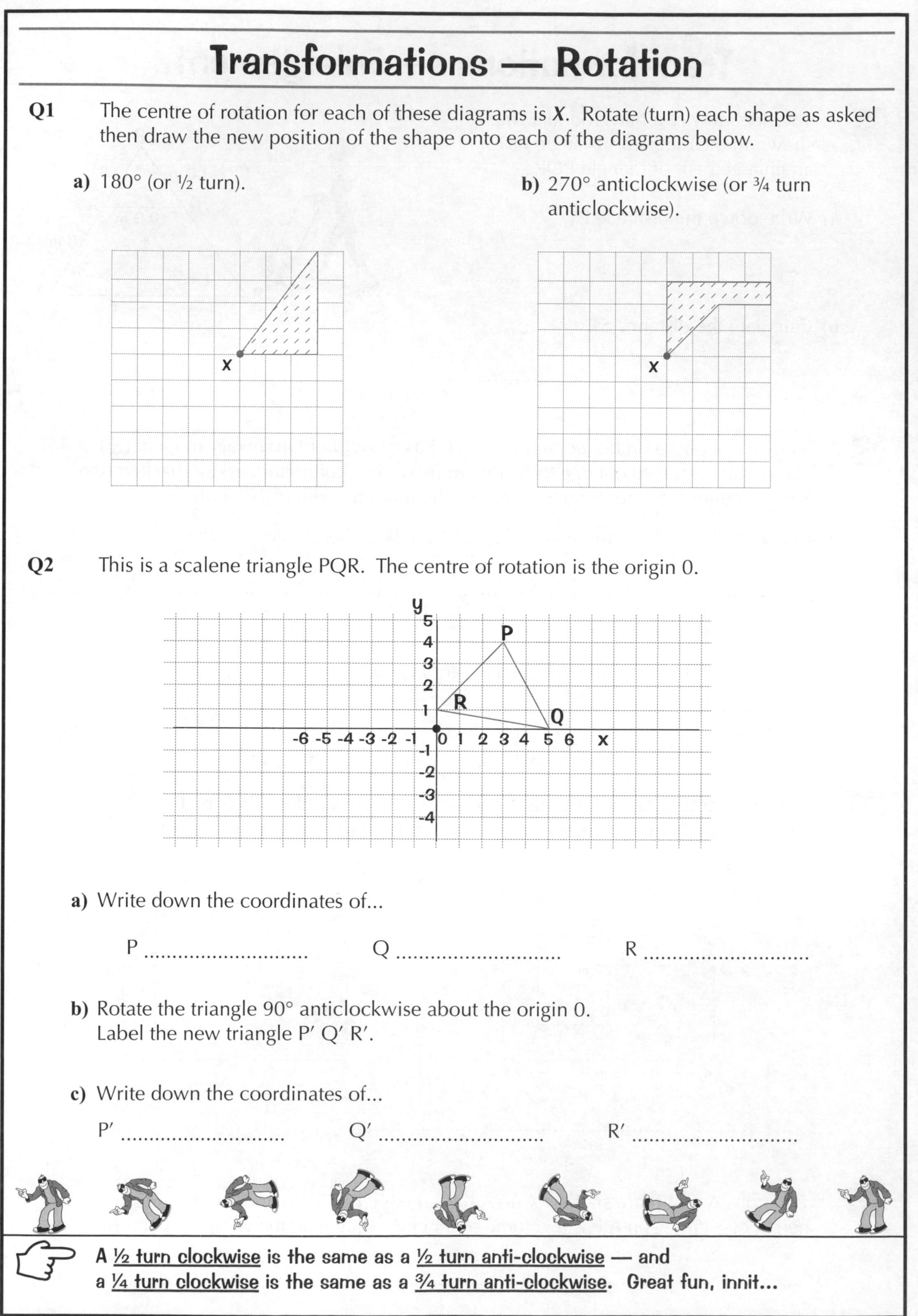

Transformations — Rotation

Q1 The centre of rotation for each of these diagrams is ***X***. Rotate (turn) each shape as asked then draw the new position of the shape onto each of the diagrams below.

a) 180° (or ½ turn).

b) 270° anticlockwise (or ¾ turn anticlockwise).

Q2 This is a scalene triangle PQR. The centre of rotation is the origin 0.

a) Write down the coordinates of...

P .... Q .... R ....

b) Rotate the triangle 90° anticlockwise about the origin 0.
Label the new triangle P′ Q′ R′.

c) Write down the coordinates of...

P′ .... Q′ .... R′ ....

A ½ turn clockwise is the same as a ½ turn anti-clockwise — and a ¼ turn clockwise is the same as a ¾ turn anti-clockwise. Great fun, innit...

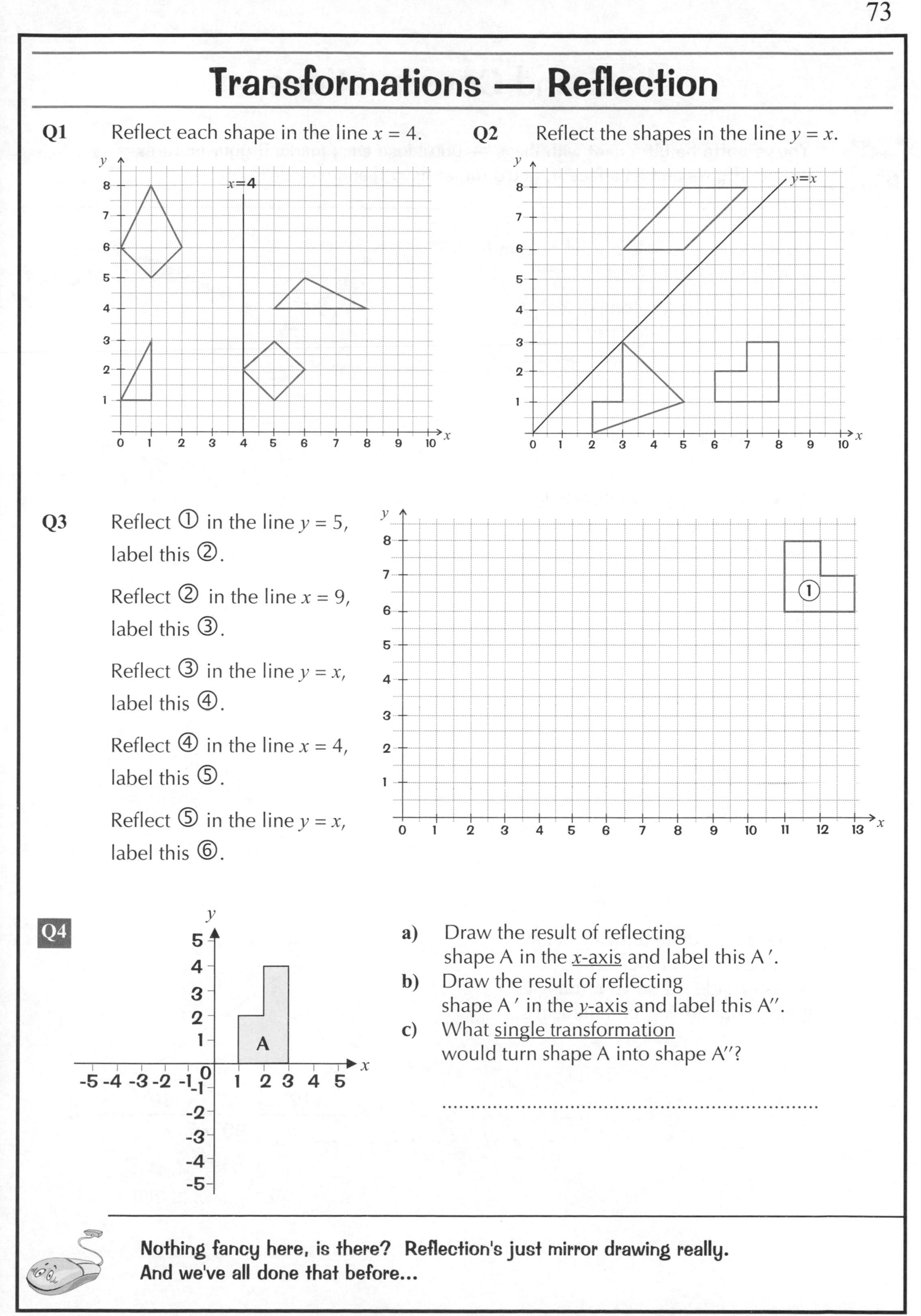
Transformations — Reflection
Q1 Reflect each shape in the line x = 4.
x=4
Q2 Reflect the shapes in the line y = x.
y=x
Q3 Reflect ① in the line y = 5, label this ②.
Reflect ② in the line x = 9, label this ③.
Reflect ③ in the line y = x, label this ④.
Reflect ④ in the line x = 4, label this ⑤.
Reflect ⑤ in the line y = x, label this ⑥.
Q4
A
a) Draw the result of reflecting shape A in the x-axis and label this A′.
b) Draw the result of reflecting shape A′ in the y-axis and label this A″.
c) What single transformation would turn shape A into shape A″?
Nothing fancy here, is there? Reflection's just mirror drawing really.
And we've all done that before...

Loci and Constructions

You've gotta be ultra neat with these — you'll lose easy marks if your pictures are scruffy — and let's face it, you'd rather have them, wouldn't you.

Constructions should always be done as accurately as possible using:

sharp pencil, ruler, compasses, protractor and set-square.

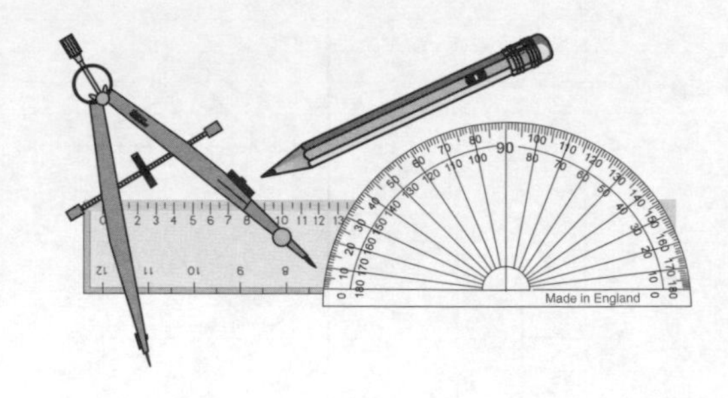

Q1 a) Draw a circle with radius 4 cm.

Draw in a diameter of the circle. Label one end of the diameter X and the other end Y.

Mark a point somewhere on the circumference — not too close to X or Y. Label your point T. Join X to T and T to Y.

Measure angle XTY.

Angle XTY = ...°

b) Make an accurate drawing below of the triangle on the right. Measure side AB on your triangle, giving your answer in millimetres.

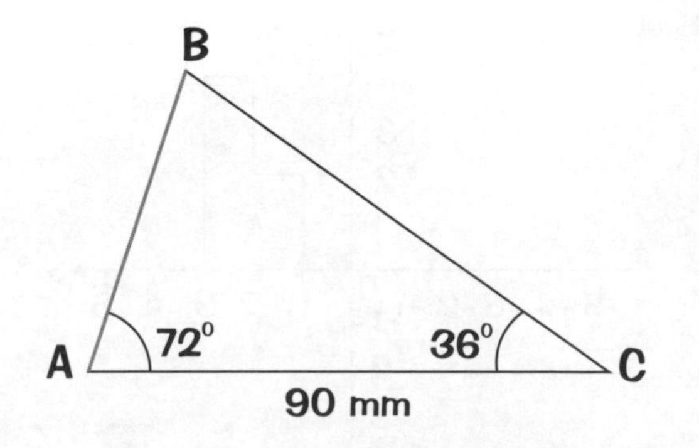

AB = ... mm

Loci and Constructions

Work through these questions bit by bit, and remember the following...

LOCUS — a line showing all points obeying the given rule.
BISECTOR — a line splitting an angle or line exactly in two.

Q2 a) In the space to the left, construct a triangle ABC with AB = 4 cm, BC = 5 cm, AC = 3 cm.

b) Construct the perpendicular bisector of AB and where this line meets BC, label the new point D.

Q3

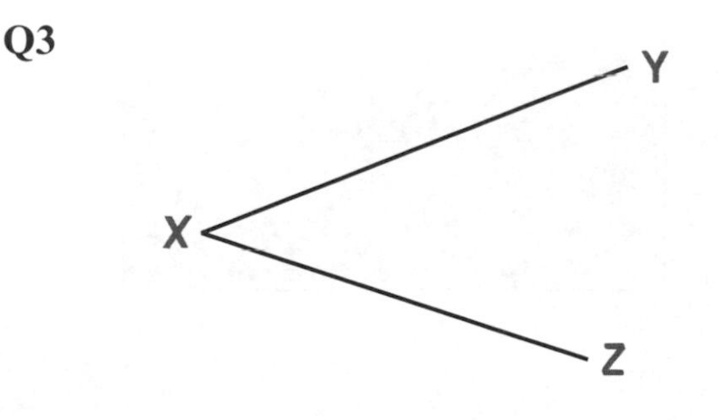

The diagram on the left shows two lines XY and XZ which meet at the point X. Construct the angle bisector of YXZ.

Q4 On a plain piece of paper mark two points A and B which are 6 cm apart.

a) Draw the locus of points which are 4 cm from A.

b) Draw the locus of points which are 3 cm from B.

c) There are 2 points which are both 4 cm from A and 3 cm from B. Label them X and Y.

Loci and Constructions

Q5 A planner is trying to decide where to build a house. He is using the plan on the right, which shows the location of a canal, a TV mast **(A)** and a water main **(B)**.
The house must be more than 100 m away from both the canal and the TV mast and be within 60 m of the water main.
Shade the area on the map where the planner could build the house.

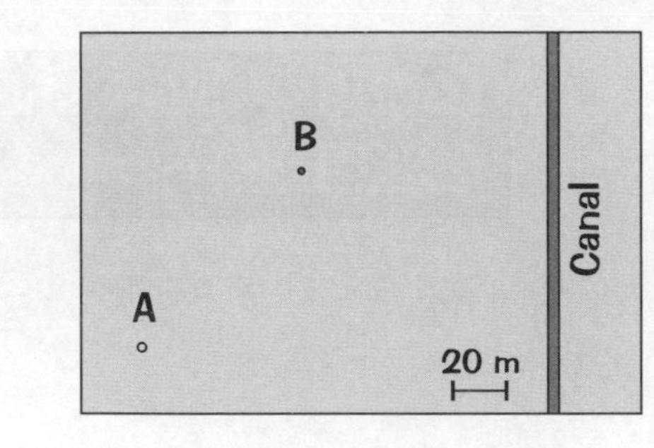

Q6 Tony likes to look at the tree in his garden. The diagram to the right shows the position of the tree relative to his bedroom window. Tony wants to position his bed in such a way that he can see the tree in the morning as he awakes.

Carefully <u>shade</u> on the diagram the area in which Tony could position his bed.

He doesn't need to be able to see the whole tree.

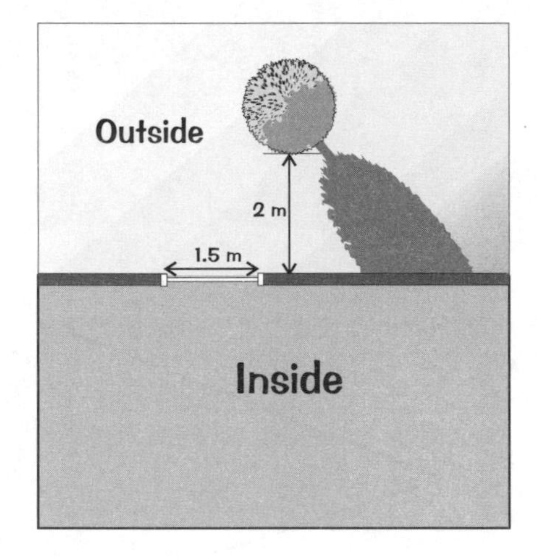

Q7 With the aid of a pair of compasses accurately draw an equilateral triangle with sides 5 cm.
Now accurately draw a square with sides 6 cm.

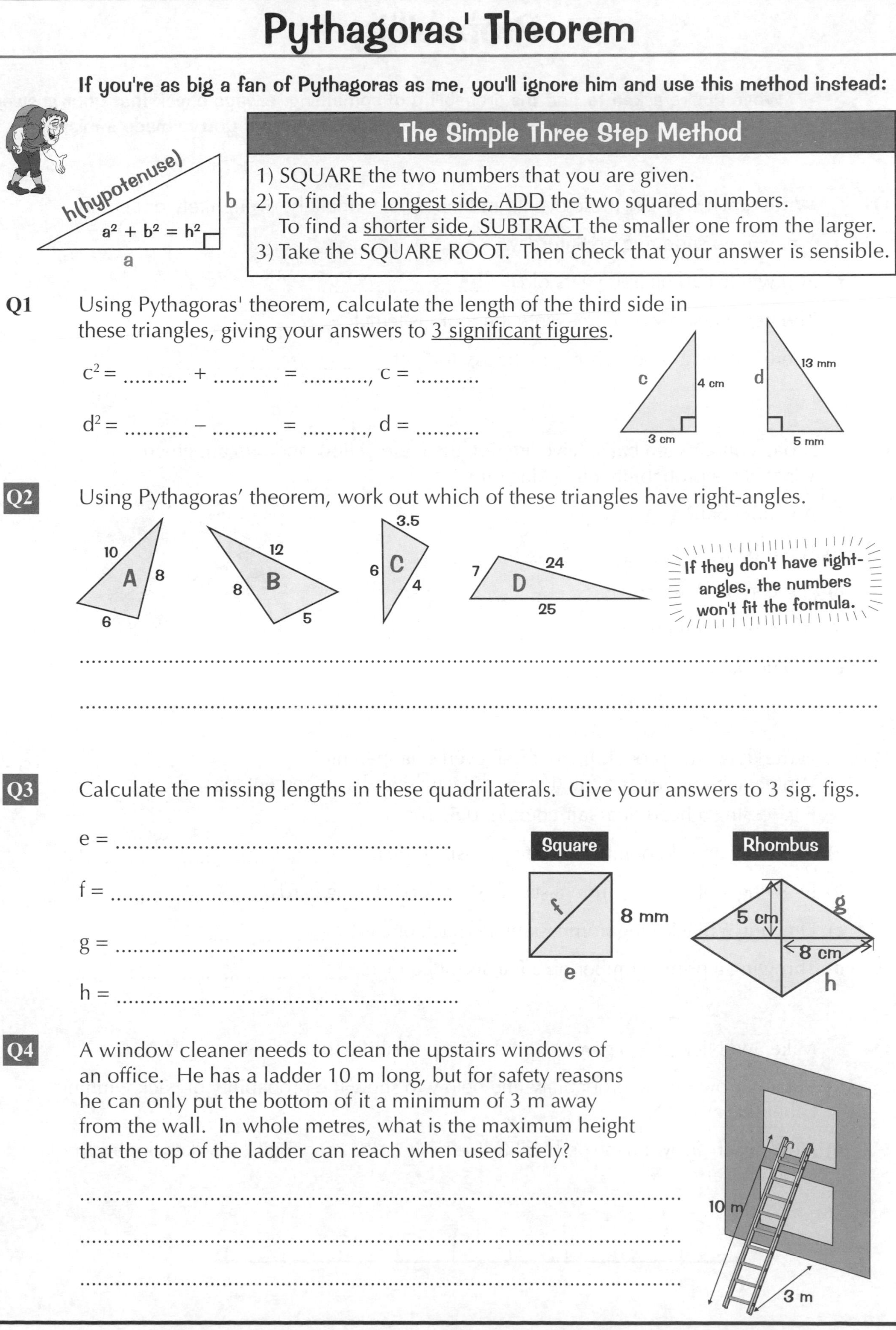

Pythagoras' Theorem

If you're as big a fan of Pythagoras as me, you'll ignore him and use this method instead:

The Simple Three Step Method

1) SQUARE the two numbers that you are given.
2) To find the longest side, ADD the two squared numbers.
   To find a shorter side, SUBTRACT the smaller one from the larger.
3) Take the SQUARE ROOT. Then check that your answer is sensible.

Q1 Using Pythagoras' theorem, calculate the length of the third side in these triangles, giving your answers to 3 significant figures.

c^2 = ... + ... = ..., c = ...

d^2 = ... − .... = ..., d = ...

Q2 Using Pythagoras' theorem, work out which of these triangles have right-angles.

....

....

Q3 Calculate the missing lengths in these quadrilaterals. Give your answers to 3 sig. figs.

e = ...

f = ...

g = ...

h = ...

Q4 A window cleaner needs to clean the upstairs windows of an office. He has a ladder 10 m long, but for safety reasons he can only put the bottom of it a minimum of 3 m away from the wall. In whole metres, what is the maximum height that the top of the ladder can reach when used safely?

....

....

....

Probability

When you're asked to find the probability of something, always check that your answer is between 0 and 1. If it's not, you know straight away that you've made a mistake.

Q1 Write down whether these events are impossible, unlikely, even, likely or certain.

a) You will go shopping on Saturday. ....

b) You will live to be 300 years old. ....

c) The next person who comes into the room is female. ....

d) There will be a moon visible in the sky tonight. ....

Q2 A bag contains ten balls. Five are red, three are yellow and two are green. What is the probability of picking out:

a) A yellow ball? ....

b) A red ball? ....

c) A green ball? ....

d) A red or a green ball? ....

e) A blue ball? ....

Q3 Write down the probability of these events happening.
Write each answer as a fraction, as a decimal and as a percentage.
E.g. tossing a head on a fair coin: ½, 0.5, 50%

a) Throwing an odd number on a fair six-sided dice. ...., ..., ...

b) Drawing a black card from a standard pack of playing cards. ...., ..., ...

c) Drawing a black King from a standard pack of cards. ...., ..., ...

d) Throwing a prime number on a fair six-sided dice. ...., ..., ...

Q4 Mike and Nick play a game of pool. The probability of Nick winning is 7/10.

a) Put an arrow on the probability line below to show the probability of Nick winning. Label this arrow N.

b) Now put an arrow on the probability line to show the probability of Mike winning the game. Label this arrow M.

0 1

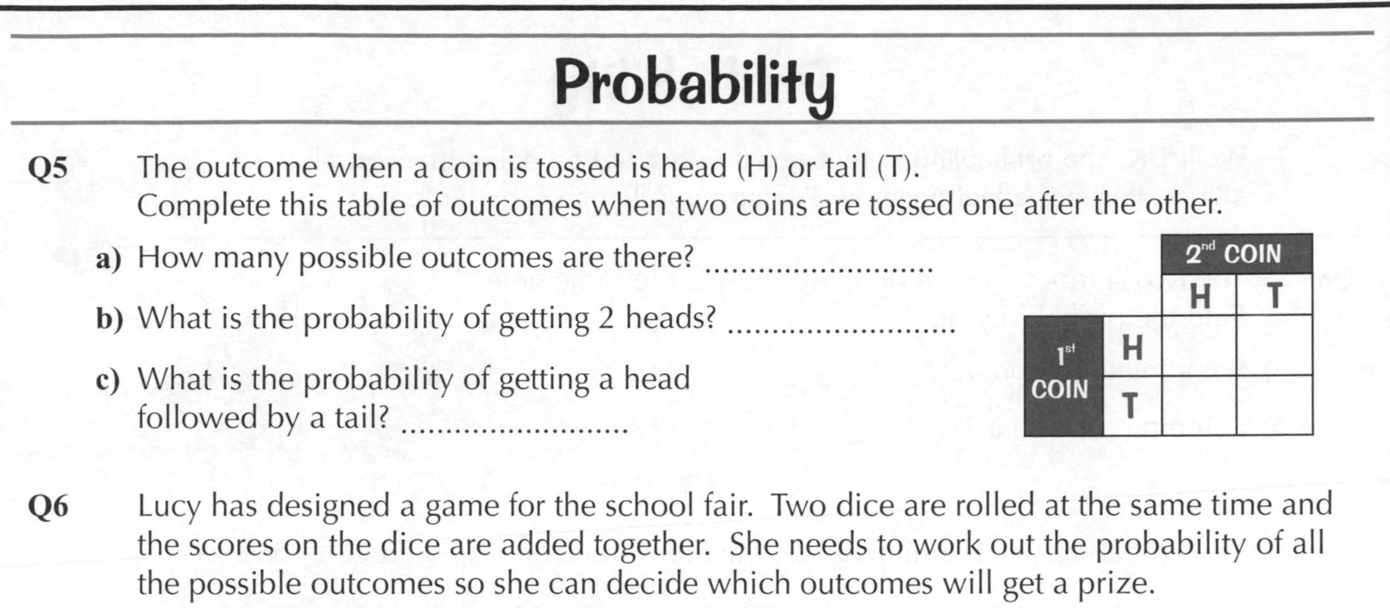

Probability

Q5 The outcome when a coin is tossed is head (H) or tail (T).
Complete this table of outcomes when two coins are tossed one after the other.

a) How many possible outcomes are there? ....

b) What is the probability of getting 2 heads? ....

c) What is the probability of getting a head followed by a tail? ....

		2nd COIN	
		H	T
1st COIN	H		
	T		

Q6 Lucy has designed a game for the school fair. Two dice are rolled at the same time and the scores on the dice are added together. She needs to work out the probability of all the possible outcomes so she can decide which outcomes will get a prize.
Complete the table of possible outcomes below.

a) How many different combinations are there? ....

		SECOND DICE					
		1	2	3	4	5	6
FIRST DICE	1						
	2	3					
	3						
	4						
	5			8			
	6						

What is the probability of scoring:

b) 2 ....

c) 6 ....

d) 10 ....

e) More than 9 ....

f) Less than 4 ....

g) More than 12 ....

h) Lucy would like to give a prize for any outcome that has a 50% chance of occurring.
Suggest an outcome that has a 50% chance of occurring.

....

Q7 Two spinners are spun and the scores are multiplied together.

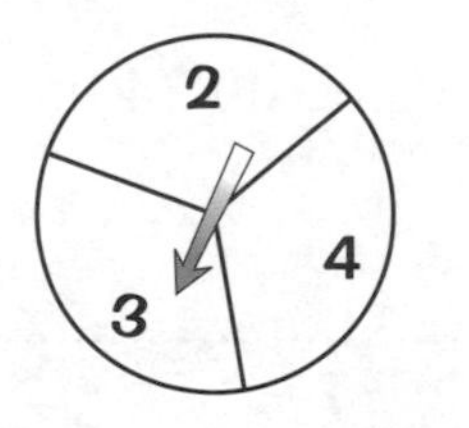

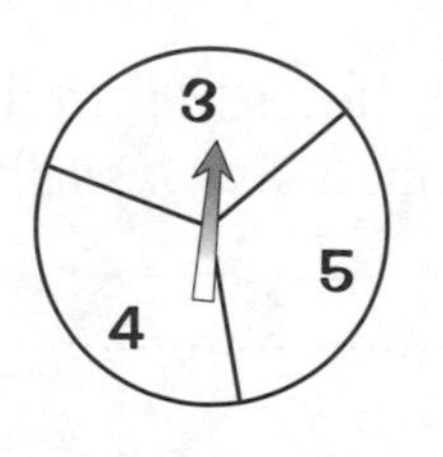

		SPINNER 1		
		2	3	4
SPINNER 2	3			
	4			
	5			

Fill in this table of possible outcomes.

What is the probability of scoring 12? ....

To win you have to score 15 or more.
What is the probability of winning? ....

In the exam, you might not be asked to put the "possible outcomes" in a table. But it's a good idea to make your own table anyway — that way you don't miss any out.

Probability

Well, OK, the probability is that you'd rather not be doing these at all... still — this is the last page, so I'm sure you'll cope for a bit longer.

Q8 The two spinners on the right are spun at the same time.
Find the probability of:

a) Spinning an A and a 2. ....

b) Spinning a 1 and a 3. ....

C B A 1 3 2

Q9 The probability of it raining during the monsoon is ¾, on a particular day.

a) What is the probability of it not raining? ....

b) If a monsoon 'season' lasts approximately 100 days, how many days are likely to be dry?

....

Q10 Imagine you have just made a 6-sided spinner in Design and Technology.
How could you test whether or not it's a fair spinner?

Remember — if the spinner's fair, the probability of landing on each side is the same.

....

....

....

Q11 **a)** A biased dice is rolled 40 times. A six came up 14 times.
Calculate the relative frequency that a six was rolled.

....

b) The same dice is rolled another 60 times. From this, a six came up 24 times.
Calculate the relative frequency that a six was rolled.

....

c) Use the data from **a)** and **b)** to make the best estimate you can of the probability of rolling a six with the dice.

....

Q12 How many times must you roll a fair 6-sided dice for the probability of getting at least one 6 to be more than 0.5?

....

Don't forget the "at least" trick — P(at least 1 six) = 1 – P(no sixes).

Data

Q1 Write down one question you could use to find out if students' exam timetables affect how well they do in exams.

....

Q2 Students at a school are doing a project on recycling in their local area. They decide to ask this question:

Are recycling levels higher since the council set up a collection service?

To answer the question, the students look at the council's data on recycling levels before and after the service was set up.

Are the students using primary or secondary data? ....

Q3 A beauty salon is promoting a new anti-wrinkle cream. They use statistics from the cosmetic company to claim that the cream is the most effective on the market.

Remember — primary data you collect yourself; secondary data is collected by someone else.

a) Is the beauty salon using primary or secondary data? ....

b) Give one disadvantage of using the cosmetic company's statistics.

....

....

Q4 Complete the table by saying which type of data has been used.

Data	Discrete or Continuous
Number of people passing through Heathrow airport each day.	
Heights of people in a Maths class.	
Number of goals scored in a hockey game.	
Time taken to be served in a fast-food restaurant.	

Don't forget: if you can measure a quantity exactly then it's discrete — otherwise it's continuous.

Data

If you're doing stats it's easy to have loads and loads of data, which can make it pretty hard to analyse. So it helps to group it together so it's more manageable.

Q5 Fred asked each of his 30 classmates how long (in minutes) it took them to eat their dinner. Here are the results he recorded:

42 13 6 31 15 20 19 5 50 14
8 25 16 27 4 45 32 31 31 10
32 17 16 19 29 42 43 30 29 18

a) Group the data appropriately and fill in the table. The first one has been done for you.

Length of time (mins)	1-10				
Number of people	5				

b) State one disadvantage of grouping data.

....

Q6 Say what the <u>population</u> is for each of these surveys:

a) The health effects of smoking on 20- to 30-year-old women.

....

b) The number of trees in public parks in London.

....

c) The pay of football players in the Premier League.

....

Q7 Cheapeez supermarket wants to find out what they should stock to attract more customers. They interview the first 100 people to go into their store one Saturday morning.

a) Give one reason why this sample will produce biased data.

....

b) What population should have been sampled from?

....

Data

Questionnaires are a great way to gather information. But you need to think carefully about how you design, distribute and collect them so you get the information you want.

Q8 Stanley is researching the use of the school canteen.
He asks this question to a sample of students at the school:
How often do you use the canteen? Tick one of the boxes.

Very often ☐ *Quite often* ☐ *Not very often* ☐

a) Give one criticism of Stanley's question.

....

b) Write a better question that Stanley could use to find out how often students at his school use the canteen.

....

Q9 A cafe owner is trying to find out which drinks his customers prefer.
He asks them to complete a questionnaire including this question:
What is your favourite drink?

i) *Tea* *ii)* *Coffee*
iii) *Fruit squash* *iv)* *Other*

52 of the first 100 people to answer reply "other".

a) Give one criticism of the question.

....

b) How could the question be improved?

....

Q10 Peter compiles a questionnaire on music tastes and sends it to a sample of 100 students at his school. He only receives 55 questionnaires back.

a) State one problem that Peter could have with his data.

....

b) What could Peter do to avoid this problem?

....

Mode and Median

To find the mode, put the data in order of size first — then it's easier to see which number you've got most of.

Q1 Find the mode for each of these sets of data.

a) 3, 5, 8, 6, 3, 7, 3, 5, 3, 9

.... Mode is ....

b) 52, 26, 13, 52, 31, 12, 26, 13, 52, 87, 41

.... Mode is ....

Q2 The midday temperatures in °C on 10 summer days in England were:

25, 18, 23, 19, 23, 24, 23, 18, 20, 19

What was the modal midday temperature?

.... Modal midday temperature was ....°C.

Q3 The time it takes thirty pupils in a class to get to school each day in minutes is:

18, 24, 12, 28, 17, 34, 17, 17, 28, 12, 23, 24, 17, 34, 9,
32, 15, 31, 17, 19, 17, 32, 15, 17, 21, 29, 34, 17, 12, 17

What is the modal time?

....

....

Modal time is .... mins.

Put the data in order of size for median questions too — it's much easier to find the middle value.

Q4 Find the median for these sets of data.

a) 3, 6, 7, 12, 2, 5, 4, 2, 9

.... Median is ....

b) 14, 5, 21, 7, 19, 3, 12, 2, 5

.... Median is ....

Q5 These are the heights of fifteen 16 year olds.

162 cm 156 cm 174 cm 148 cm 152 cm 139 cm 167 cm 134 cm
157 cm 163 cm 149 cm 134 cm 158 cm 172 cm 146 cm

What is the median height? Write your answer in the shaded box.

															Median

Mean and Range

Yikes — MEAN questions... well, they're not as bad as everyone makes out. Remember to include zeros in your calculations — they still count.

Q1 Find the mean of each of the sets of data below. If necessary, round your answers to 1 decimal place:

a) 13, 15, 11, 12, 16, 13, 11, 9 = ...

b) 16, 13, 2, 15, 0, 9 = ...

c) 80, 70, 80, 50, 60, 70, 90, 60, 50, 70, 70 = ...

Remember the formula for mean — total of the items ÷ number of items.

Q2 Find, <u>without</u> a calculator, the mean for each of these sets of data:

a) 5, 3, 7, 3, 2 = ...

b) 7, 3, 9, 5, 3, 5, 4, 6, 2, 6 = ...

Q3 The number of goals scored by a hockey team over a period of 10 games is listed below.

0, 3, 2, 4, 1, 2, 3, 4, 1, 0.

What is the range of the number of goals scored? ....

Q4 Sarah and her friends were measured and their heights were found to be:

1.52 m, 1.61 m, 1.49 m, 1.55 m, 1.39 m, 1.56 m.

What is the range of the heights? ....

Q5 Here are the times 6 people took to do a Maths test:

1 hour 10 mins, 2 hours 10 mins, 1 hour 35 mins,
1 hours 55 min, 1 hour 18 mins, 2 hours 15 mins.

What is the range of these times? ....

Q6 David is on a diet. He needs to have a mean calorie intake of 1900 calories a day. The table below shows how many calories he has eaten so far this week.

How many calories must David eat on Friday to make sure that his mean intake for the five days is correct?

....

Day	Calories Eaten
Monday	1875
Tuesday	2105
Wednesday	1680
Thursday	1910
Friday	

Averages

Geesh — as if it's not enough to make you work out all these boring averages, they want you to write stuff about them as well. Oh well, here goes nothing.

Q1 Here are the marks John and Mark got on their last five maths tests.

John	65	83	58	79	75
Mark	72	70	81	67	70

Calculate the range for each pupil.
Who has the most consistent results? Explain your answer.

....

....

Q2 The bar graph shows the amount of time Jim and Bob spend watching TV during the week.

a) Find the mean amount of time per day each spends watching TV.

....

....

b) Find the range of times for each of them.

....

c) Using your answers from **a)** and **b)**, comment on what you notice about the way they watch TV.

....

Q3 The Borders Orchid Growers Society has measured the heights of all the Lesser Plumed Bog Orchids in the 5 mile wide strip each side of the English-Scottish border. Their heights are given below to the nearest cm.

Scottish side
Heights 14, 15, 17, 14, 17, 16, 14, 13
15, 17, 16, 14, 15, 17, 14, 13

English side
Heights 14, 12, 16, 18, 19, 17, 16, 15
13, 14, 15, 16, 17, 18, 19, 13

a) State the mode and median for each set of data.

....

b) Find the range for each set of data.

....

c) On which side of the border are you likely to see taller orchids? Explain your answer.

....

Averages

Q4 On a large box of matches it says "Average contents 240 matches".
I counted the number of matches in ten boxes. These are the results:

241 244 236 240 239 242 237 239 239 236

Is the label on the box correct? Use the mean, median and mode for the numbers of matches to explain your answer.

....

....

Q5 The shoe sizes in a class of girls are:

3 3 4 4 5 5 5 5 6 6 6 7 8

a) Calculate the mean, median and mode for the shoe sizes.

....

....

b) If you were a shoe shop manager, which average would be most useful to you, and why?

....

Q6 A house-building company needs a bricklayer.
This advert appears in the local newspaper.
The company employs the following people:

Bricklayer wanted
Average wage over £350 p.w.

Position	Wage (p.w.)
Director	£700
Foreman	£360
Plasterer	£300
Bricklayer	£250
Bricklayer	£250

a) What is the median wage?

....

b) What is the mean wage? ....

c) Which gives the best idea of the average wage? ....

d) Is the advert fair? Explain your answer.

....

....

Write a fairer advert in the space above.

Frequency Tables

Q1 At the British Motor Show 60 people were asked what type of car they preferred. Jeremy wrote down their replies using a simple letter code.

Saloon - S Hatchback - H 4 × 4 - F MPV - M Roadster - R

Here is the full list of replies.

H	S	R	S	S	R	M	F	S	S	R	R
M	H	S	H	R	H	M	S	F	S	M	S
R	R	H	H	H	S	M	S	S	R	H	H
H	H	R	R	S	S	M	M	R	H	M	H
H	S	R	F	F	R	F	S	M	S	H	F

a) Fill in the table and add up the frequency in each row.

TYPE OF CAR	TALLY	FREQUENCY
Saloon		
Hatchback		
4 × 4		
MPV		
Roadster		

b) Draw a frequency graph of the results.

A frequency graph just means a bar chart.

FREQUENCY: 0, 2, 4, 6, 8, 10, 12, 14, 16, 18

TYPE OF CAR: S H F M R

c) What is the modal type of car? ....

Q2 Last season Newcaster City played 32 matches.
The number of goals they scored in each match were recorded as shown.

2	4	3	5
1	0	3	2
4	2	1	2
0	2	3	1

1	0	0	1
1	1	1	0
1	3	2	0
1	1	0	4

a) Complete the frequency table.

GOALS	TALLY	FREQUENCY	GOALS × FREQUENCY
0			
1			
2			
3			
4			
5			

b) What is the mean number of goals scored?

....

c) What is the modal number of goals scored?

....

Frequency Tables

Q3 A tornado has struck the hamlet of Moose-on-the-Wold. Many houses have had windows broken. The frequency table shows the devastating effects.

No. of windows broken per house	0	1	2	3	4	5	6	Totals
Frequency	5	3	4	11	13	7	2	
Windows × Frequency								

a) What is the modal number of broken windows? ....

b) Complete the frequency table.

c) Find the median number of broken windows.

....

d) Calculate the mean number of broken windows.

....

e) State the range of the data ....

Q4 The frequency table below shows the number of hours spent Christmas shopping by 100 people surveyed in a town centre.

Number of Hours	0	1	2	3	4	5	6	7	8
Frequency	1	9	10	10	11	27	9	15	8
Hours × Frequency									

a) What is the modal number of hours spent Christmas shopping? ....

b) Fill in the third row of the table.

c) What is the total amount of time spent Christmas shopping by all the people surveyed?

....

d) What is the mean amount of time spent Christmas shopping by a person?

....

Grouped Frequency Tables

As a rule these are trickier than standard frequency tables — you'll certainly have to tread carefully here. Have a good look at the box below and make sure you remember it.

Mean and Mid-Interval Values

1) The **MID-INTERVAL VALUES** are just what they sound like — the middle of the group.
2) Using the Frequencies and Mid-Interval Values you can estimate the **MEAN**.

$$\text{Estimated Mean} = \frac{\text{Overall Total (Frequency} \times \text{Mid-interval value)}}{\text{Frequency total}}$$

Shoe Size	1 - 2	3 - 4	5 - 6	7 - 8	Totals
Frequency	15	10	3	1	29
Mid-Interval Value	1.5	3.5	5.5	7.5	—
Frequency x Mid-Interval Value	22.5	35	16.5	7.5	81.5

So the estimated mean value is 81.5 ÷ 29 = 2.81

Q1 In a survey of test results in a French class at Blugdon High, these grades were achieved by the 23 pupils:

(grade) score	(E) 31-40	(D) 41-50	(C) 51-60	(B) 61-70
frequency	4	7	8	4

a) Write down the mid-interval values for each of the groups.

....

b) Calculate an estimate for the mean value.

....

....

Q2 Dean is carrying out a survey for his Geography coursework. He asked 80 people how many miles they drove their car last year to the nearest thousand. He has started filling in a grouped frequency table to show his results.

No. of Miles (thousands)	1 - 10	11 - 20	21 - 30	31 - 40	41 - 50	51 - 60	61 - 70	71 - 80	81 - 90	91 - 100
No. of Cars	2	3	5	19	16	14	10	7		

a) Complete Dean's table using the following information:

81 245, 82 675, 90 159, 90 569

b) Write down the modal class. ....

c) Which class contains the median number of miles driven? ....

Grouped Frequency Tables

Q3 This table shows times for two teams of swimmers, the Dolphins and the Sharks.

Dolphins			Sharks		
Time interval (seconds)	Frequency	Mid-interval value	Time interval (seconds)	Frequency	Mid-interval value
$14 \le t < 20$	3	17	$14 \le t < 20$	6	17
$20 \le t < 26$	7	23	$20 \le t < 26$	15	23
$26 \le t < 32$	15		$26 \le t < 32$	33	
$32 \le t < 38$	32		$32 \le t < 38$	59	
$38 \le t < 44$	45		$38 \le t < 44$	20	
$44 \le t < 50$	30		$44 \le t < 50$	8	
$50 \le t < 56$	5		$50 \le t < 56$	2	

a) Complete the table, writing in all mid-interval values.

b) Use the mid-interval technique to estimate the mean time for each team.

....

....

Q4 The lengths of 25 snakes are measured to the nearest cm, and then grouped in a frequency table.

Length	151 - 155	156 - 160	161 -165	166 - 170	171 - 175	Total
frequency	4	8	7	5	1	25

Which of the following sentences may be true and which have to be false?

a) The median length is 161 cm. ....

b) The range is 20 cm. ....

c) The modal class has 7 snakes.

....

d) The median length is 158 cm.

....

Remember — with grouped data you can't find these values exactly, but you can narrow down the possibilities.

Tables, Charts and Graphs

Make sure you read these questions carefully. You don't want to lose easy marks by looking at the wrong bit of the table or chart.

Q1 The local library is carrying out a survey to find the amount of time people spend reading each day. Complete the frequency table below, then draw a frequency polygon to show the results.

Time spent reading (mins)	Tally	Frequency
1 - 15	~~||||~~ |	6
16 - 30	~~||||~~ |||	
31 - 45	|||	
46 - 60	~~||||~~	
61 - 75	|||	

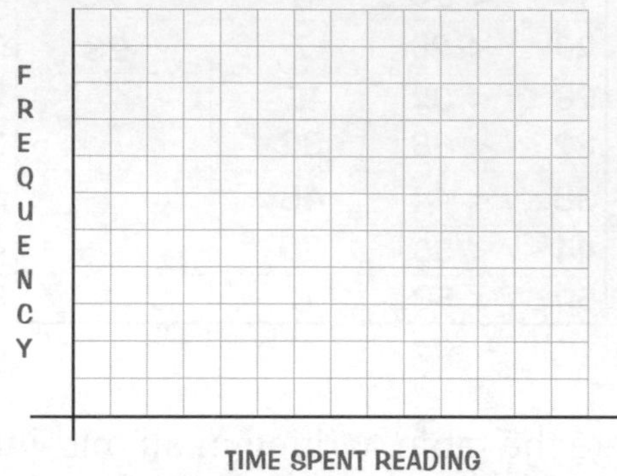

Q2 One hundred vehicles on a road were recorded as part of a traffic study. Use this two-way table to answer the following questions.

	Van	Motor-bike	Car	Total
Travelling North	15			48
Travelling South	20		23	
Total		21		100

a) How many vans were recorded? ....

b) How many vehicles in the survey were travelling south? ....

c) How many motorbikes were travelling south? ....

d) How many cars were travelling north? ....

Q3 This pictogram shows the favourite drinks of a group of pupils.

Favourite Drinks	Number of Pupils
Lemonade	✧ ✧ ✧ ✧ ✧ ✧ ✧ ✧ ✧
Cola	✧ ✧ ✧ ✧ ✧ ✧ ✧ ✧ ✧ ✧ ✧
Cherryade	✧ ✧ ✧ ✧ ✧ ✧
Orange Squash	✧ ✧ ✧
Milk	✧

✧ Represents 2 pupils.

a) How many pupils were questioned? .... pupils.

b) How many pupils prefer non-fizzy drinks? .... pupils.

c) 18 pupils liked lemonade best. How many more liked cola best? .... pupils.

d) Comment on the popularity of cola compared with milk.

....

Tables, Charts and Graphs

Q4 This stem and leaf diagram shows the ages of people in a cinema.

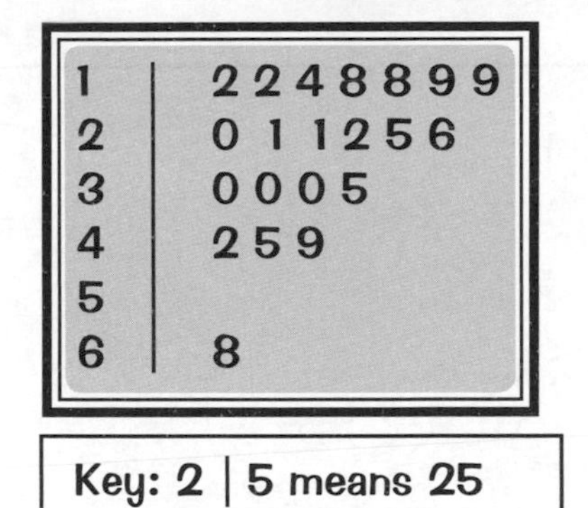

1	2 2 4 8 8 9 9
2	0 1 1 2 5 6
3	0 0 0 5
4	2 5 9
5	
6	8

Key: 2 | 5 means 25

a) How many people in the cinema were in their twenties? ....

b) Write out the ages of all of the people in the cinema below.

....

Q5 This stem and leaf diagram shows the exam scores of a group of Year 9 pupils.

a) How many pupils got a score between 60 and 70? ....

b) How many scored 80 or more? ....

c) What was the most common test score? ....

d) How many scored less than 50? ....

e) How many pupils took the test? ....

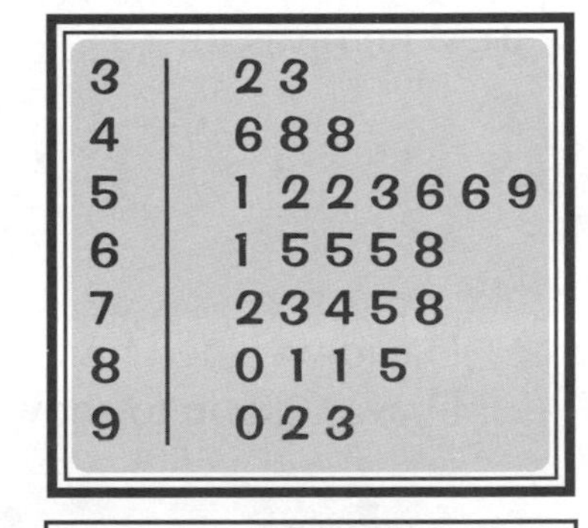

3	2 3
4	6 8 8
5	1 2 2 3 6 6 9
6	1 5 5 5 8
7	2 3 4 5 8
8	0 1 1 5
9	0 2 3

Key: 5 | 2 means 52

Q6 I've been measuring the length of beetles for a science project. Here are the lengths in millimetres:

12 18 20 11 31
19 27 34 19 22

a) Complete the stem and leaf diagram on the right to show the results above.

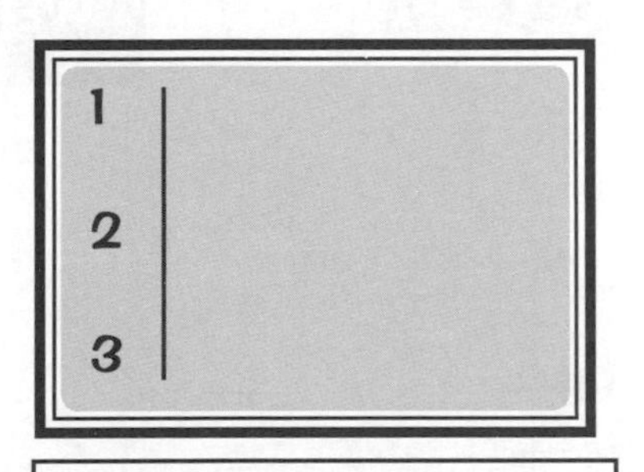

1	
2	
3	

Key: 2 | 2 means 22

b) From your stem and leaf diagram, find the range of the data.

....

c) Find the median of the data.

....

Stem and leaf diagrams are easy once you understand how they work. Just remember that the digit to the left of the line is the 'tens' digit.

Tables, Charts and Graphs

Drawing line graphs is easy — just plot the points, then join them up with straight lines.

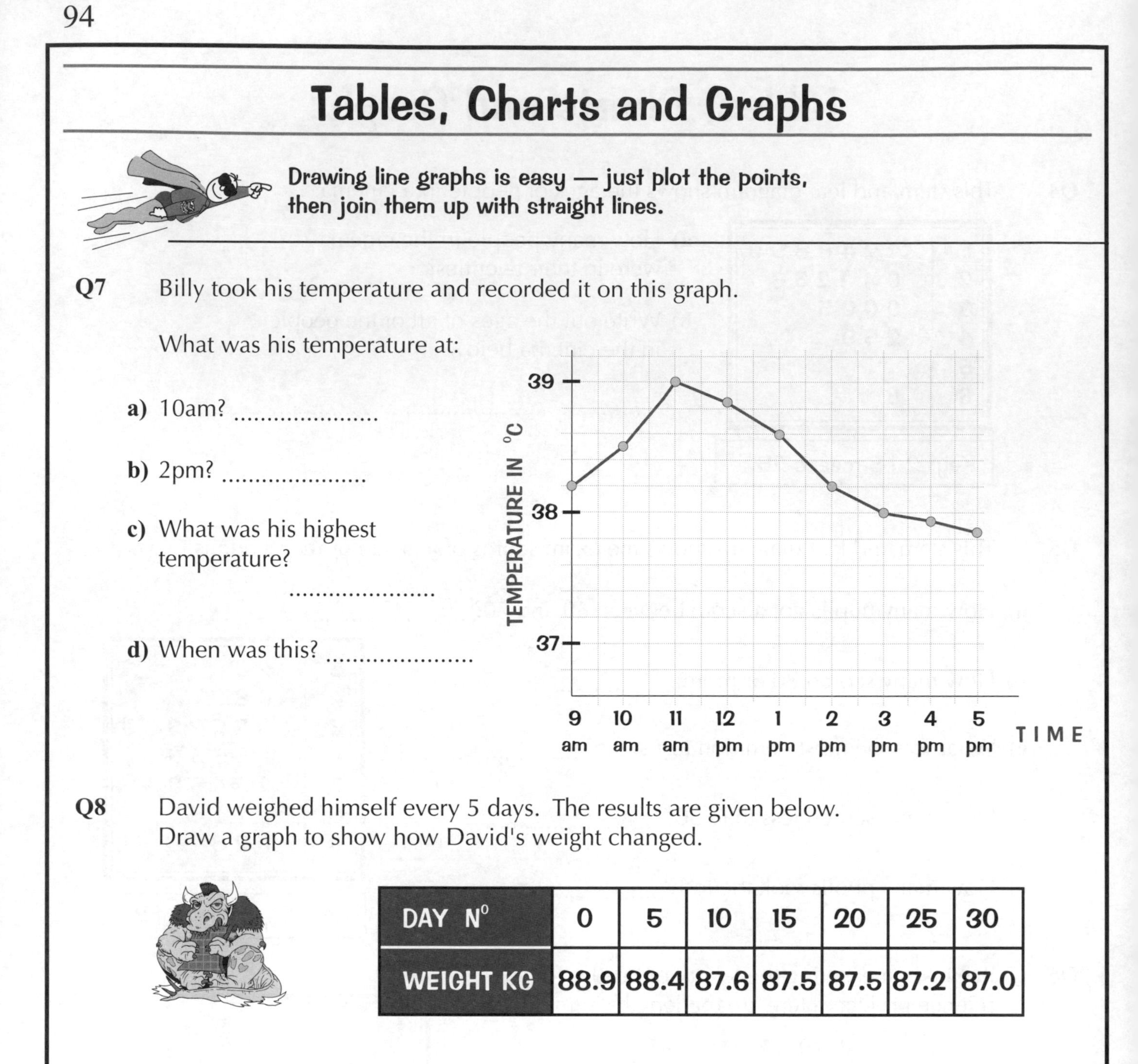

Q7 Billy took his temperature and recorded it on this graph.

What was his temperature at:

a) 10am? ....

b) 2pm? ....

c) What was his highest temperature?

....

d) When was this? ....

Q8 David weighed himself every 5 days. The results are given below. Draw a graph to show how David's weight changed.

DAY N°	0	5	10	15	20	25	30
WEIGHT KG	88.9	88.4	87.6	87.5	87.5	87.2	87.0

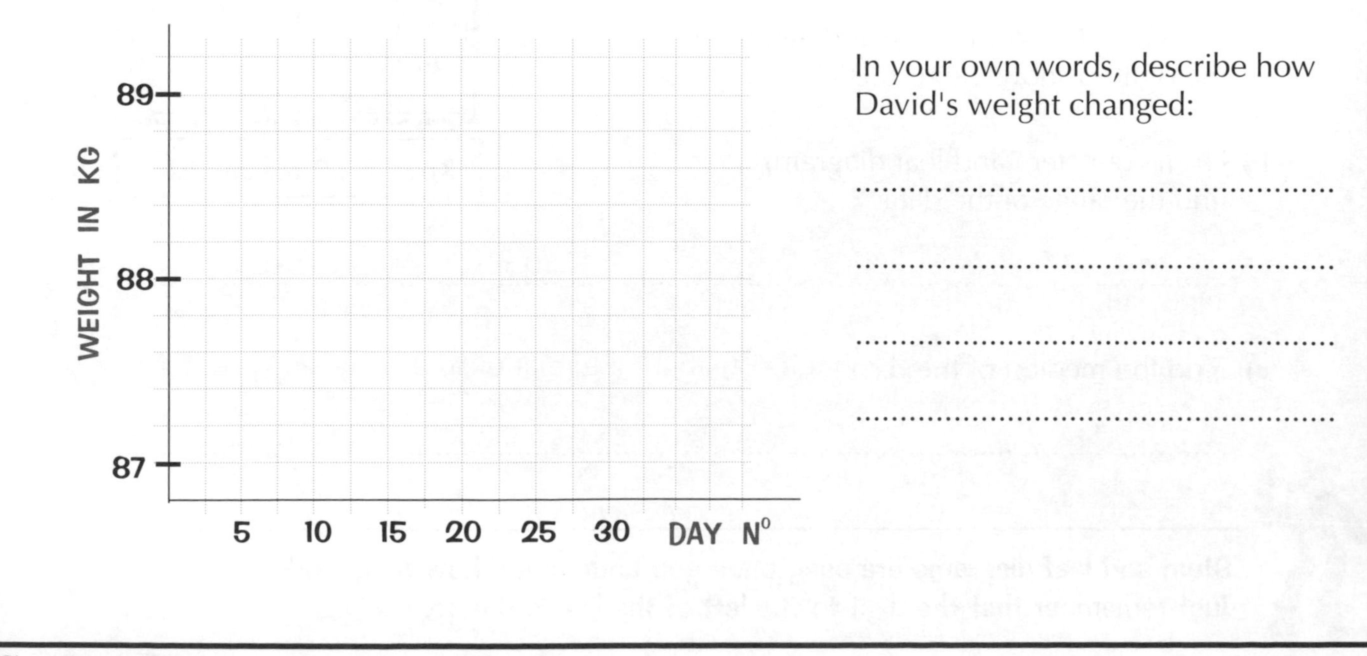

In your own words, describe how David's weight changed:

....

....

....

....

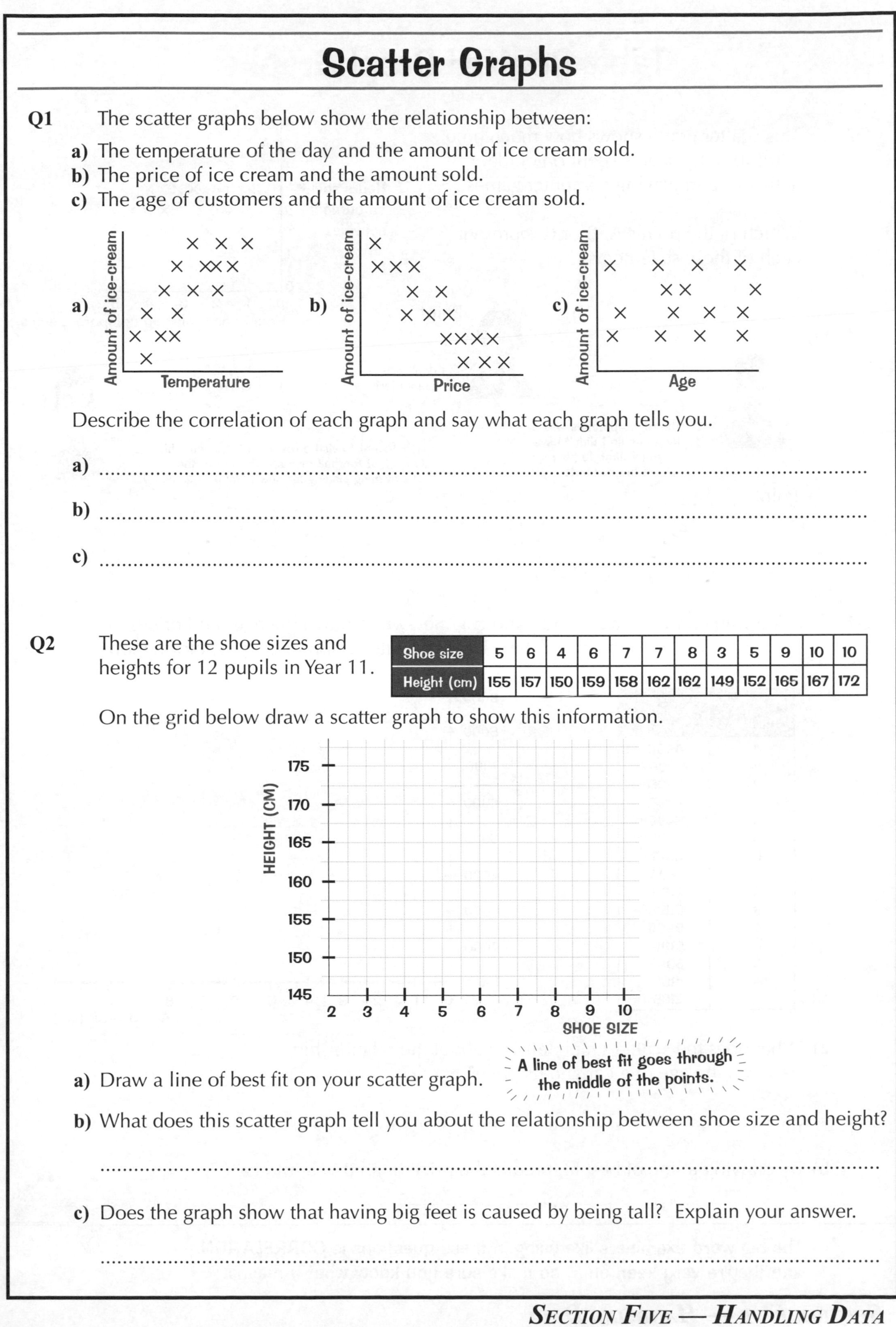

Scatter Graphs

Q1 The scatter graphs below show the relationship between:

a) The temperature of the day and the amount of ice cream sold.
b) The price of ice cream and the amount sold.
c) The age of customers and the amount of ice cream sold.

Describe the correlation of each graph and say what each graph tells you.

a) ....

b) ....

c) ....

Q2 These are the shoe sizes and heights for 12 pupils in Year 11.

Shoe size	5	6	4	6	7	7	8	3	5	9	10	10
Height (cm)	155	157	150	159	158	162	162	149	152	165	167	172

On the grid below draw a scatter graph to show this information.

a) Draw a line of best fit on your scatter graph.

b) What does this scatter graph tell you about the relationship between shoe size and height?

....

c) Does the graph show that having big feet is caused by being tall? Explain your answer.

....

Scatter Graphs

Q3 This scatter graph shows how much time a group of teenagers spent on outdoor activities and playing computer games.

Which of the points A, B or C represent each of these statements?

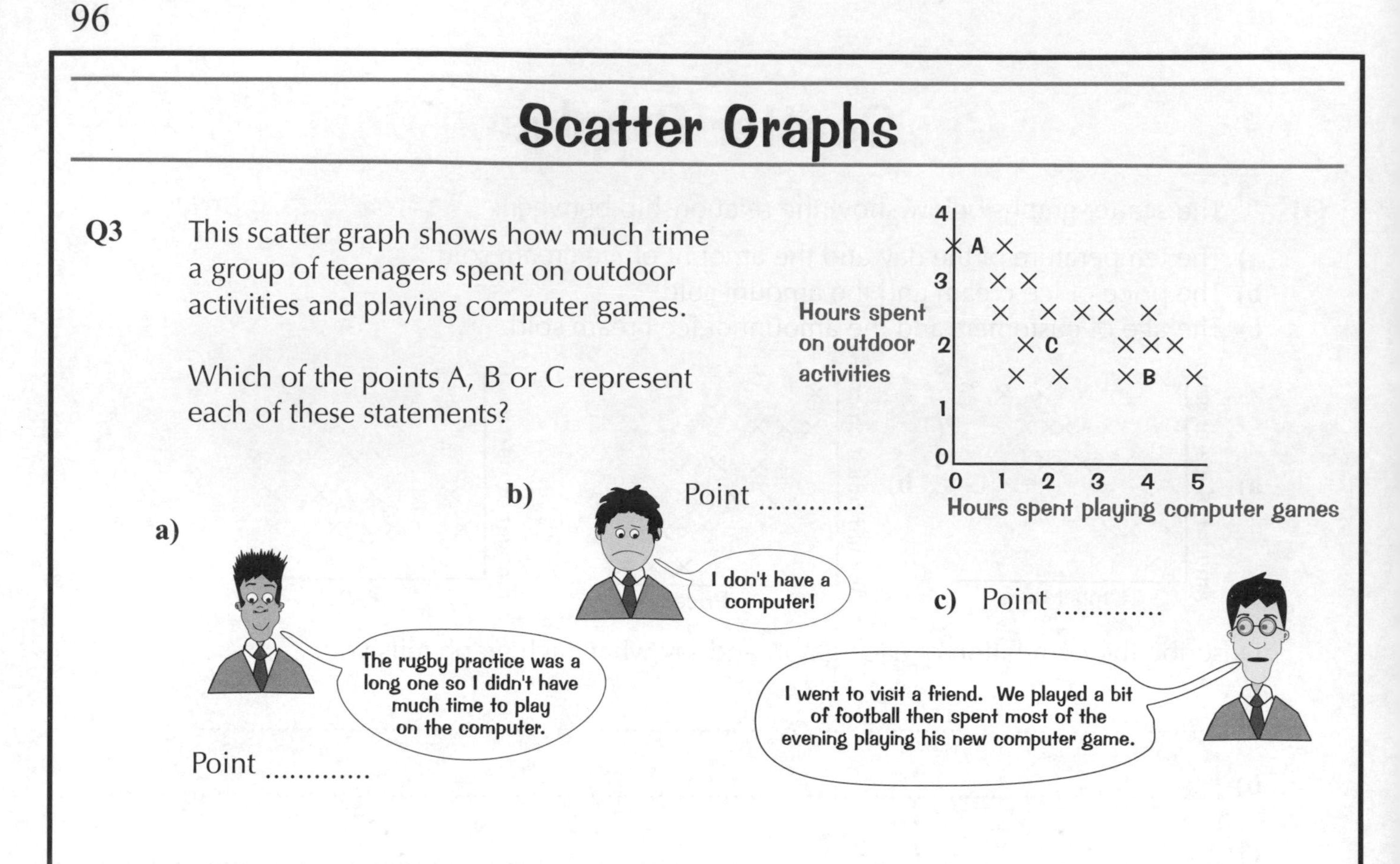

a) Point ....

b) Point ....

c) Point ....

Q4 Alice wanted to buy a second-hand car. She wrote down the ages and prices of 15 similarly-sized cars on sale locally. Draw a scatter graph for Alice's information.

Age of car (years)	Price (£)
4	4995
2	7995
3	6595
1	7995
5	3495
8	4595
9	1995
1	7695
2	7795
6	3995
5	3995
1	9195
3	5995
4	4195
9	2195

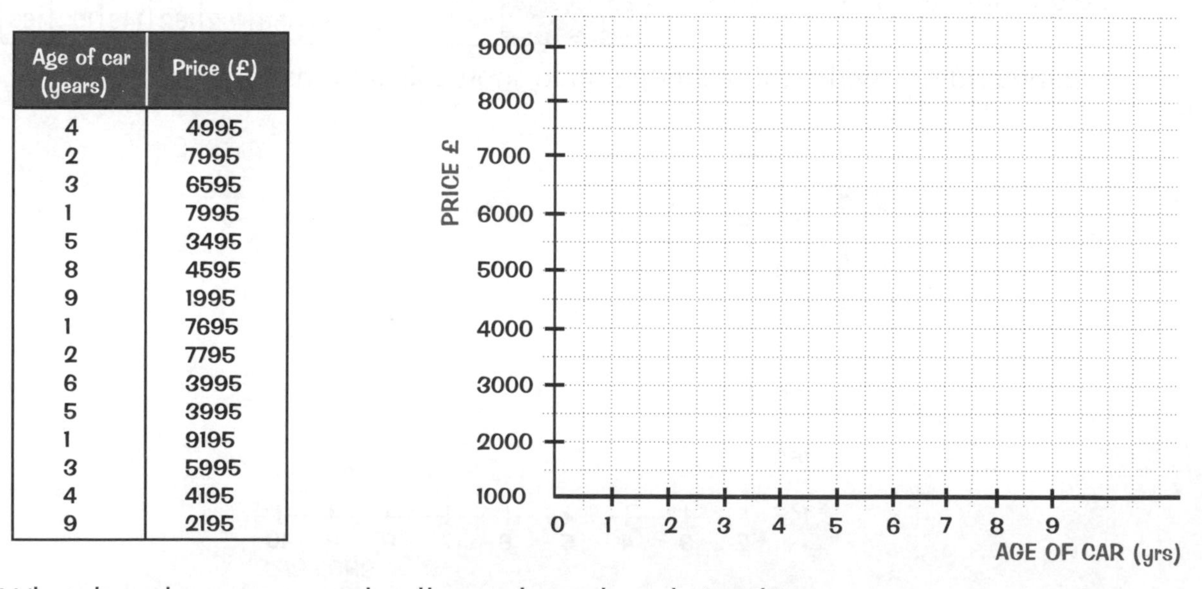

a) What does the scatter graph tell you about the relationship between the age of a car and its price?

....

b) By drawing a line of best fit, predict the price of an 8-year-old car. ....

The big word examiners like using in these questions is CORRELATION — and they're very keen on it, so make sure you know what it means.

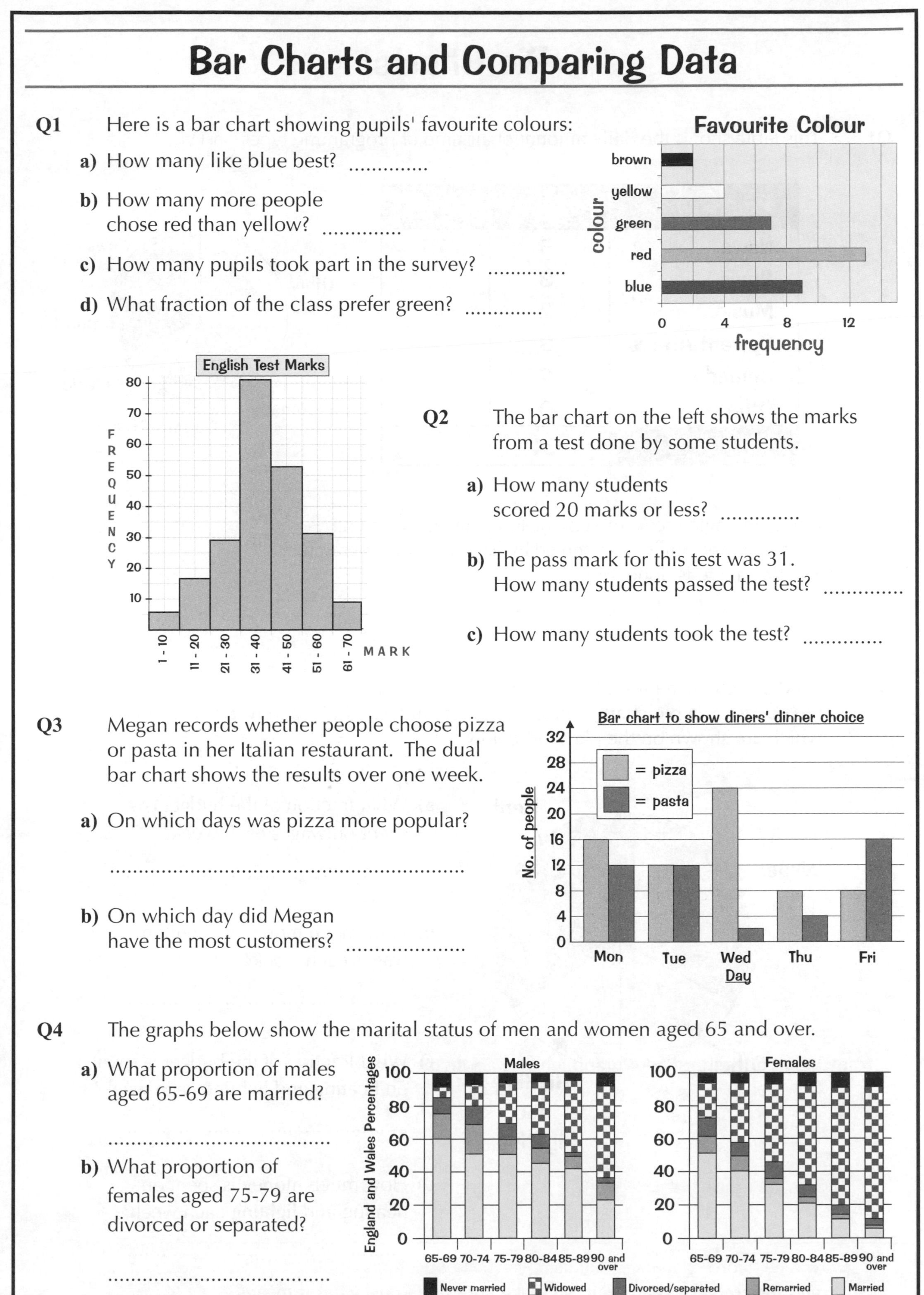

Bar Charts and Comparing Data

Q1 Here is a bar chart showing pupils' favourite colours:

a) How many like blue best? ....

b) How many more people chose red than yellow? ....

c) How many pupils took part in the survey? ....

d) What fraction of the class prefer green? ....

Q2 The bar chart on the left shows the marks from a test done by some students.

a) How many students scored 20 marks or less? ....

b) The pass mark for this test was 31. How many students passed the test? ....

c) How many students took the test? ....

Q3 Megan records whether people choose pizza or pasta in her Italian restaurant. The dual bar chart shows the results over one week.

a) On which days was pizza more popular?

....

b) On which day did Megan have the most customers? ....

Q4 The graphs below show the marital status of men and women aged 65 and over.

a) What proportion of males aged 65-69 are married?

....

b) What proportion of females aged 75-79 are divorced or separated?

....

Pie Charts

Q1 This table shows the daily amount of air time of programme types on TV:

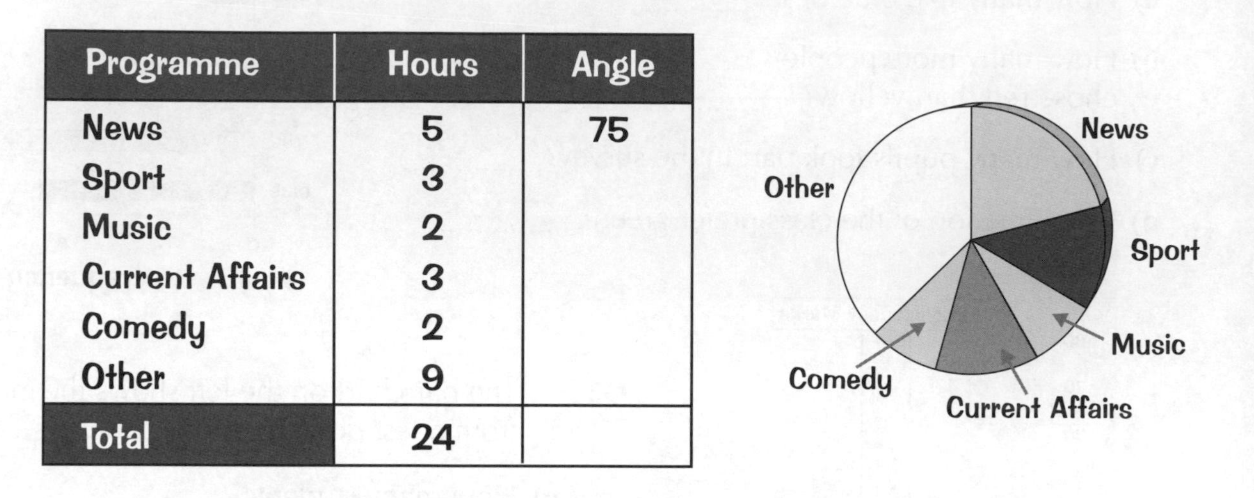

Programme	Hours	Angle
News	5	75
Sport	3	
Music	2	
Current Affairs	3	
Comedy	2	
Other	9	
Total	24	

Using an angle measurer, complete the table by finding the size of the angle represented by each type of programme. The first angle is done for you.

Q2 Sandra is giving a presentation on her company's budget. She has decided to present the budget as a pie chart. The company spends £54 000 each week on various items, which are shown on the pie chart below.

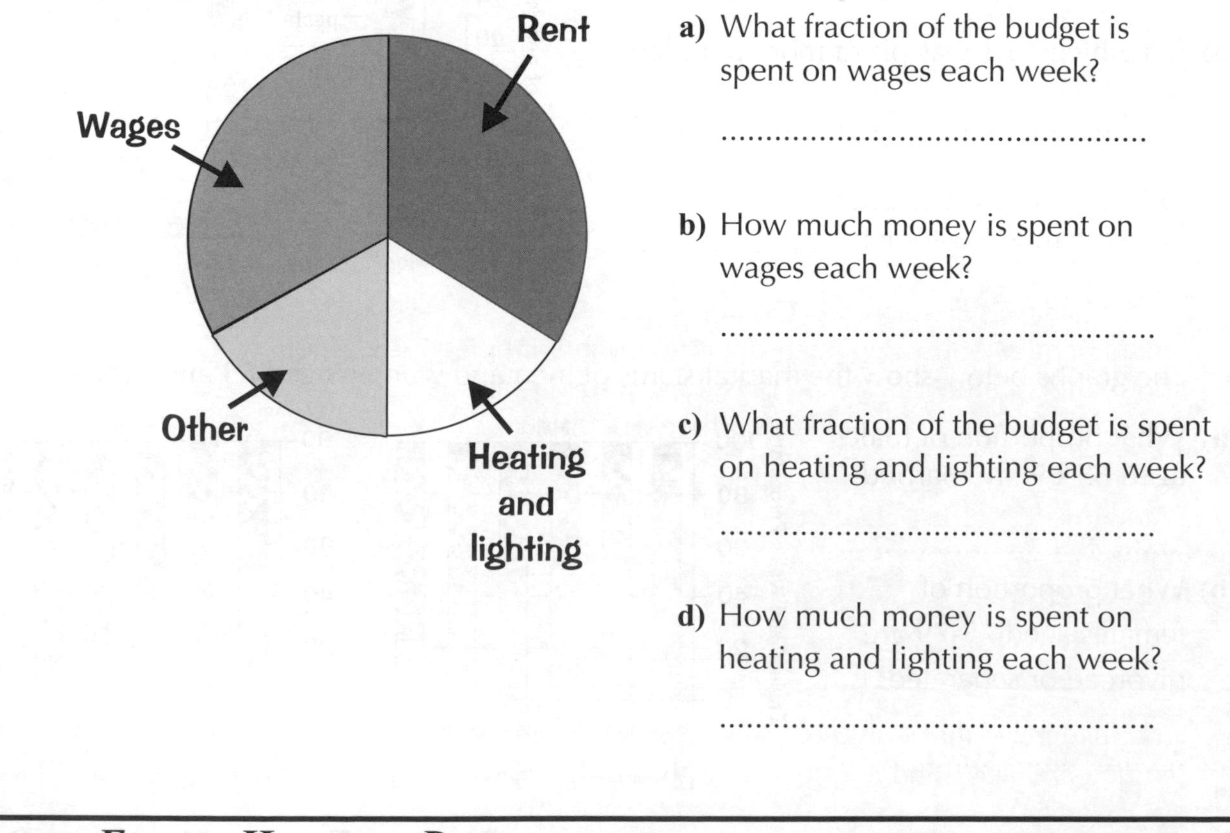

a) What fraction of the budget is spent on wages each week?

....

b) How much money is spent on wages each week?

....

c) What fraction of the budget is spent on heating and lighting each week?

....

d) How much money is spent on heating and lighting each week?

....

Pie Charts

Q3 In a University department there are 180 students from different countries.

Country	UK	Malaysia	Spain	Others
Number of students	90	35	10	45

To show this on a pie chart you have to work out the angle of each sector.
Complete the table showing your working. The UK is done for you.

COUNTRY	WORKING	ANGLE in degrees
UK	90 ÷ 180 × 360 =	180°
MALAYSIA		
SPAIN		
OTHERS		

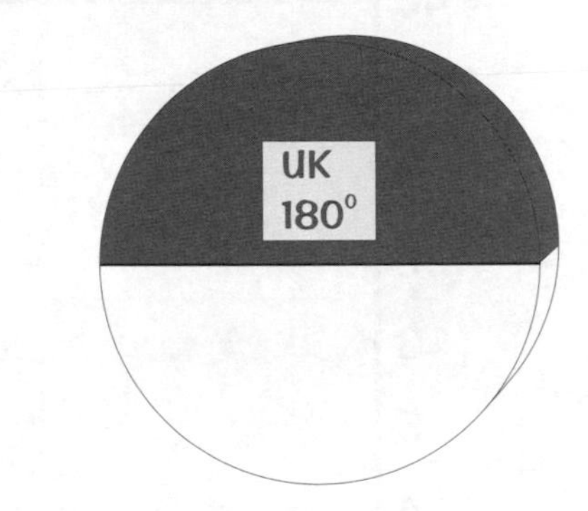

Now complete the pie chart using an angle measurer. The UK sector is done for you.

Q4 Pupils at a school were asked about their activities at weekends. The results are shown in the table. Complete the table and then draw the pie chart using an angle measurer.

ACTIVITY	HOURS	WORKING	ANGLE
Homework	6	6 ÷ 48 × 360 =	45°
Sport	2		
TV	10		
Computer games	2		
Sleeping	18		
Listening to music	2		
Paid work	8		
Total	48		

Q5 The pie charts opposite appear in a newspaper article about a local election. Nicki says that more people voted for the Green Party in 2009 than in 2005.

Comment on <u>whether it's possible</u> to tell this from the pie charts.

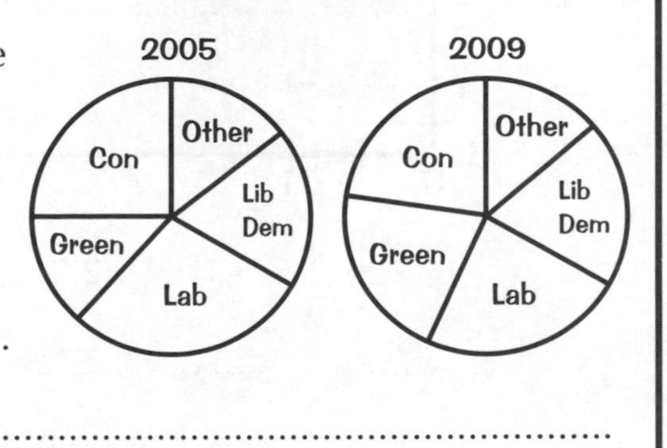

....

....

The full circle (that's all 360° of it) represents the total of everything — so you shouldn't find any gaps in it, basically.

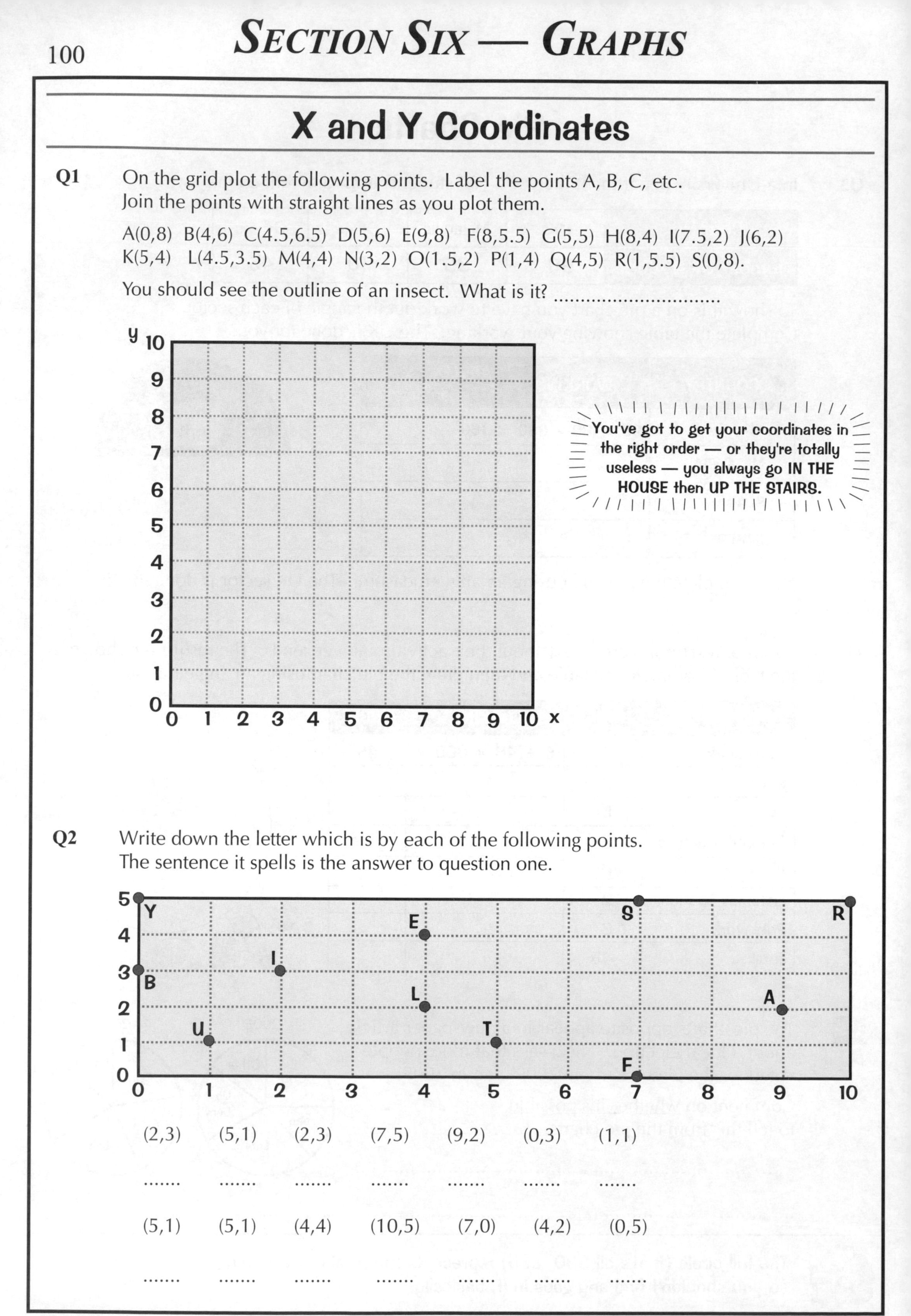

X and Y Coordinates

Q1 On the grid plot the following points. Label the points A, B, C, etc.
Join the points with straight lines as you plot them.

A(0,8) B(4,6) C(4.5,6.5) D(5,6) E(9,8) F(8,5.5) G(5,5) H(8,4) I(7.5,2) J(6,2)
K(5,4) L(4.5,3.5) M(4,4) N(3,2) O(1.5,2) P(1,4) Q(4,5) R(1,5.5) S(0,8).

You should see the outline of an insect. What is it? ....

Q2 Write down the letter which is by each of the following points.
The sentence it spells is the answer to question one.

(2,3) (5,1) (2,3) (7,5) (9,2) (0,3) (1,1)

....

(5,1) (5,1) (4,4) (10,5) (7,0) (4,2) (0,5)

....

X and Y Coordinates

Remember — 1) x comes before y

2) x goes a-cross (get it) the page. (Ah, the old ones are the best...)

Q3 The map shows the island of Tenerife where the sun never stops shining...

a) Use the map to write down the coordinates of the following:

Airport (.... , ...)

Mount Teide (.... , ...)

Santa Cruz (.... , ...)

Puerto Colon (.... , ...)

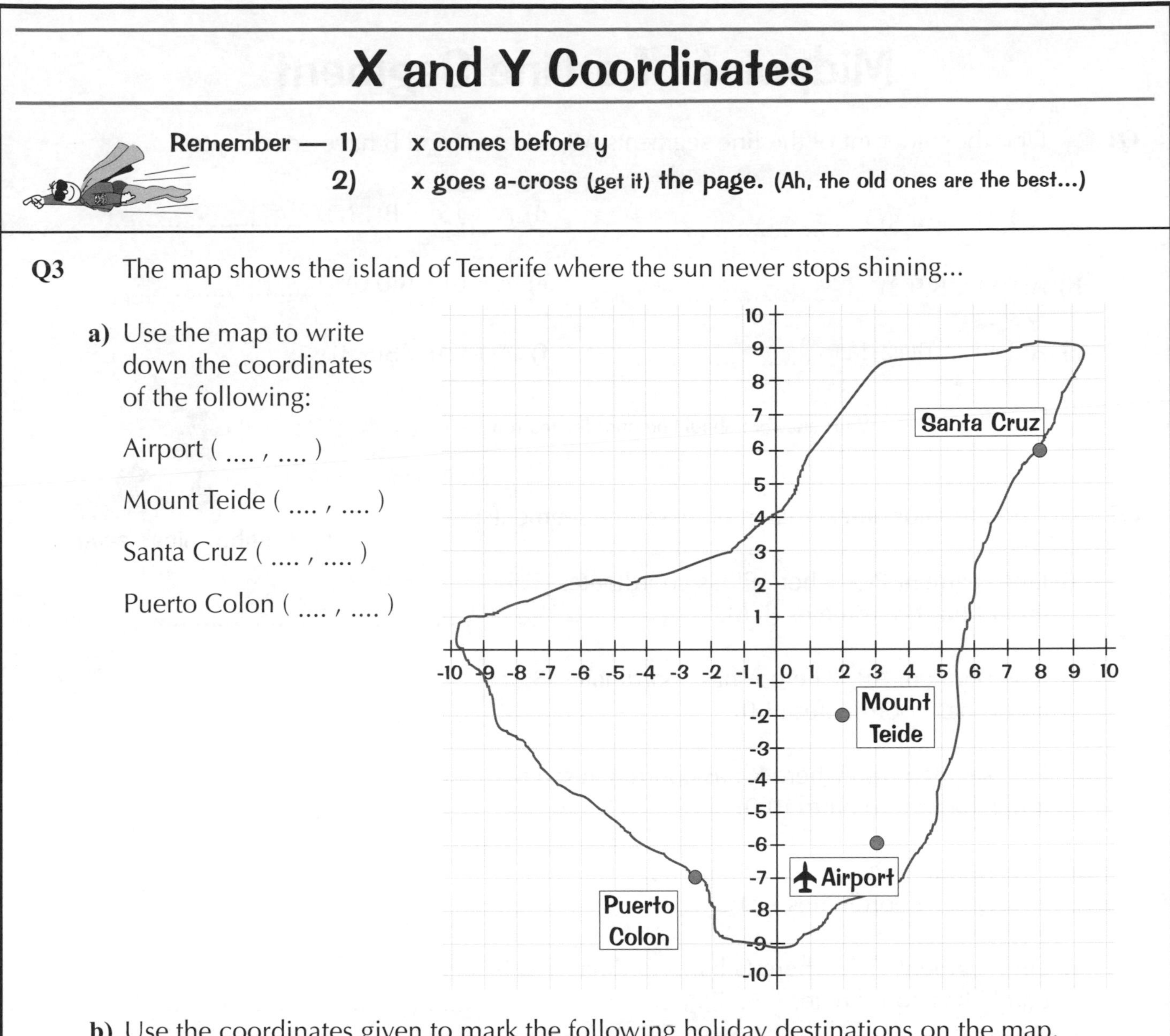

b) Use the coordinates given to mark the following holiday destinations on the map.
Las Americas (-4 , -6), El Medano (4 , -4), Icod (-6 , 2), Laguna (3 , 7), Taganana (9 , 9)

c) The cable car takes you to the top of Mount Teide. It starts at (3 , 1) and ends at (2 , -2). Draw the cable car route on the map.

Q4 On the graph paper below, draw the following lines:

a) $y = -2$

b) $x = -4$

c) $y = 1.5$

d) $x = 2.5$

Midpoint of a Line Segment

Q1 Find the midpoint of the line segments AB, where A and B have coordinates:

a) A(2,3) B(4,5) ....

b) A(1,8) B(9,2) ....

c) A(0,11) B(12,11) ....

d) A(3,15) B(13,3) ....

e) A(6,6) B(0,0) ....

f) A(15,9) B(3,3) ....

Your answers should be coordinates too.

ahh... nice'n'easy...

Q2 Find the midpoints of each of these line segments:

a) Line segment PQ, where P has coordinates (1,5) and Q has coordinates (5,6). ....

b) Line segment AB, where A has coordinates (3,3) and B has coordinates (4,0). ....

c) Line segment RS, where R has coordinates (4,5) and S has coordinates (0,0). ....

d) Line segment PQ, where P has coordinates (1,3) and Q has coordinates (3,1). ....

e) Line segment GH, where G has coordinates (0,0) and H has coordinates (–6,–7). ....

Q3 A college wants to install drinking water fountains at convenient locations around the college site (shown below on the coordinate map). The fountains are to be placed at points exactly half way between the main buildings.

a) Where should they be placed between the following buildings? Give the coordinates.

Art and Biology: ....

Gym and Humanities: ....

Languages and Maths: ....

y
4
Maths
3
Languages
2
Art
Biology
Chemistry
1
Physics
-4 -3 -2 -1 1 2 3 4 5 x
Humanities
-1
Dining Hall
-2
Finance Office
-3
Gym
Entrance
-4

b) What building is halfway between Physics and the Finance Office? Give the name and coordinates.

....

Straight-Line Graphs

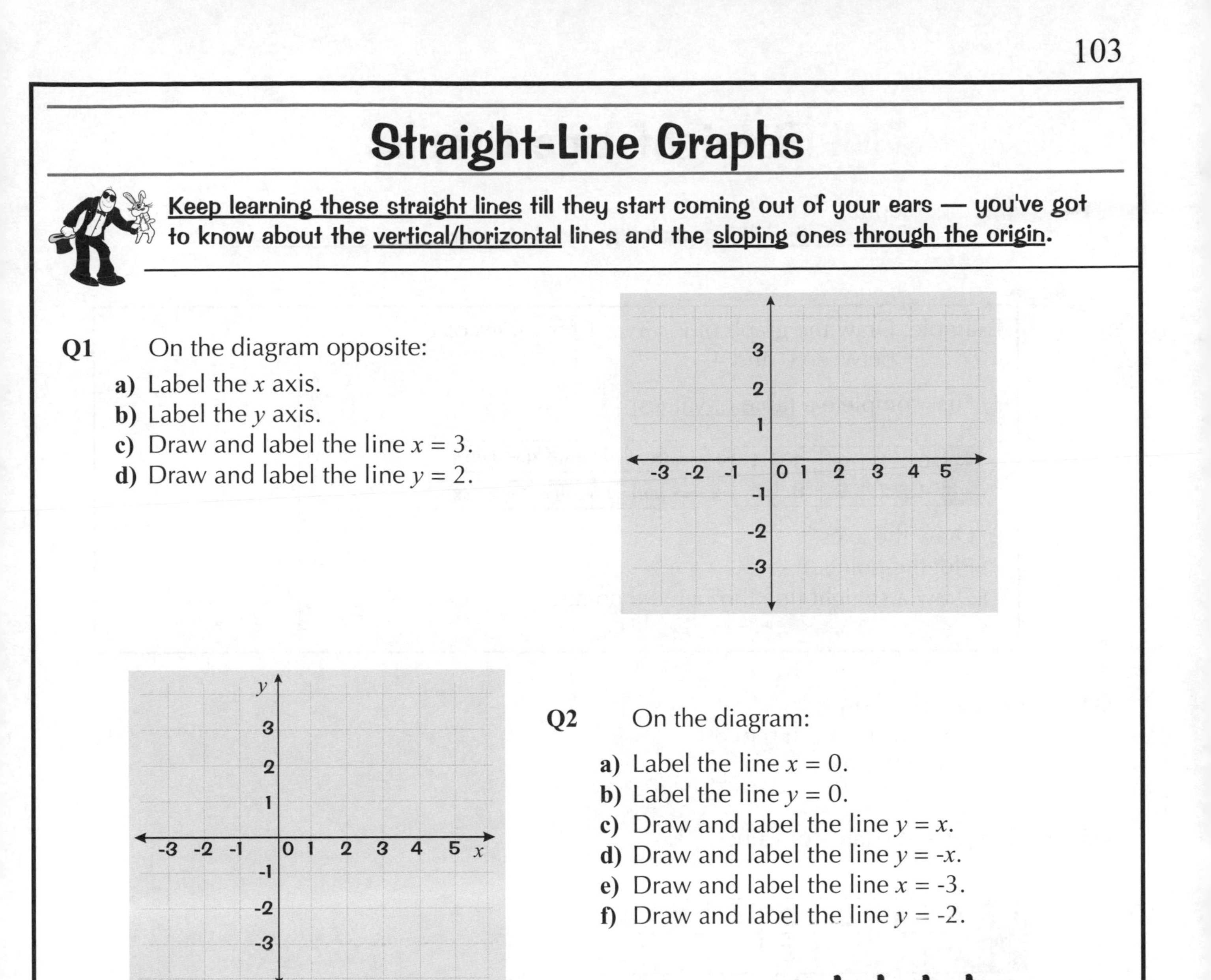

Keep learning these straight lines till they start coming out of your ears — you've got to know about the vertical/horizontal lines and the sloping ones through the origin.

Q1 On the diagram opposite:

a) Label the x axis.
b) Label the y axis.
c) Draw and label the line $x = 3$.
d) Draw and label the line $y = 2$.

Q2 On the diagram:

a) Label the line $x = 0$.
b) Label the line $y = 0$.
c) Draw and label the line $y = x$.
d) Draw and label the line $y = -x$.
e) Draw and label the line $x = -3$.
f) Draw and label the line $y = -2$.

Q3 Which letters represent the following lines:

a) $x = y$? ....

b) $x = 5$? ....

c) $y = -x$? ....

d) $x = 0$? ....

e) $y = -7$? ....

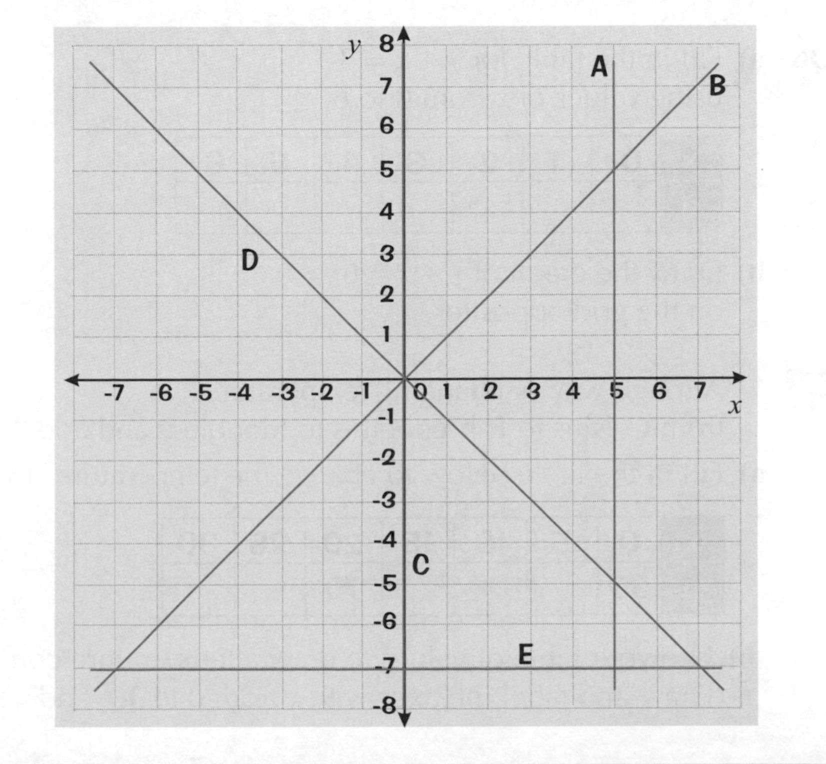

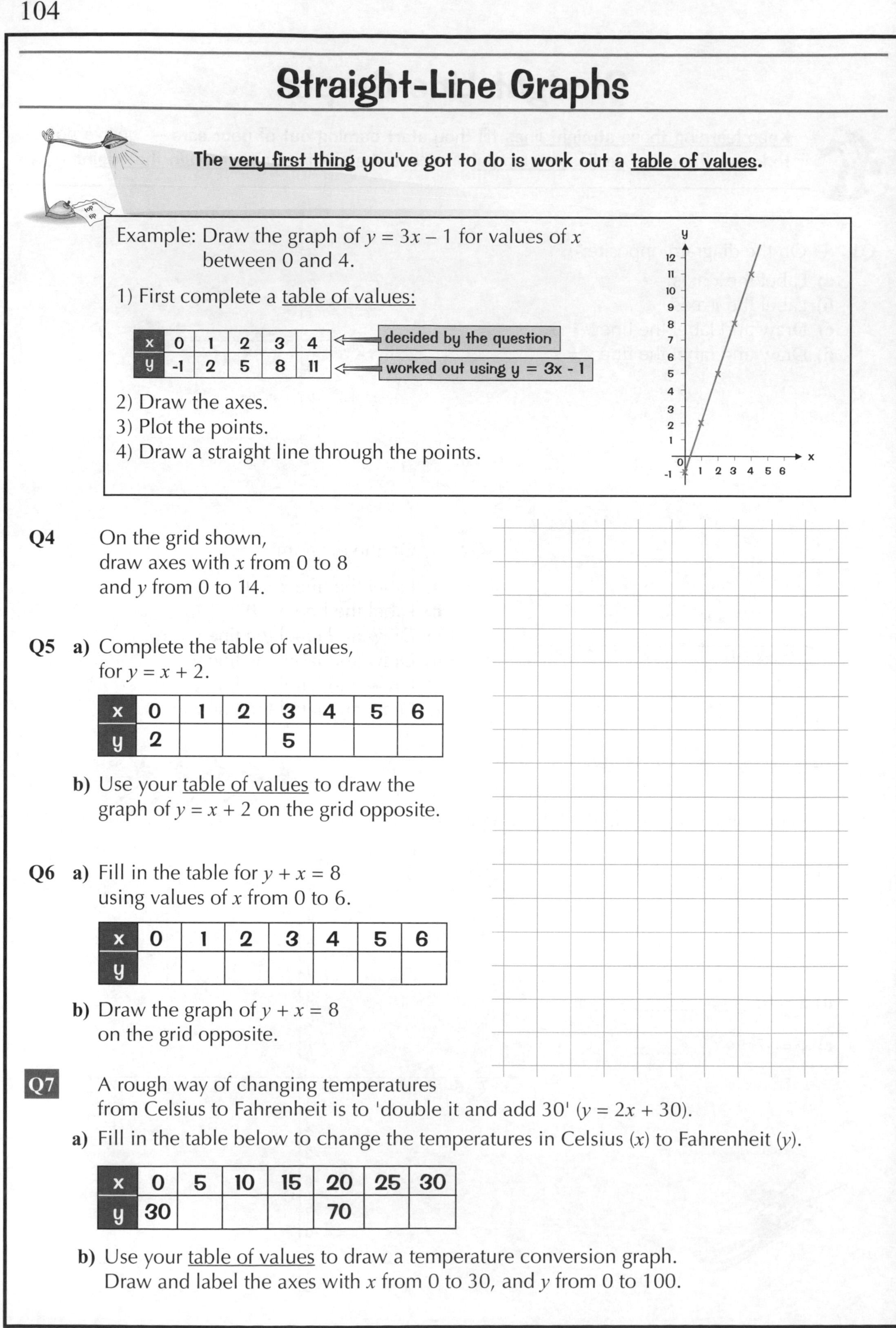

Straight-Line Graphs

The very first thing you've got to do is work out a table of values.

Example: Draw the graph of $y = 3x - 1$ for values of x between 0 and 4.

1) First complete a table of values:

x	0	1	2	3	4
y	-1	2	5	8	11

decided by the question (x row)

worked out using y = 3x - 1 (y row)

2) Draw the axes.
3) Plot the points.
4) Draw a straight line through the points.

Q4 On the grid shown, draw axes with x from 0 to 8 and y from 0 to 14.

Q5 a) Complete the table of values, for $y = x + 2$.

x	0	1	2	3	4	5	6
y	2			5			

b) Use your table of values to draw the graph of $y = x + 2$ on the grid opposite.

Q6 a) Fill in the table for $y + x = 8$ using values of x from 0 to 6.

x	0	1	2	3	4	5	6
y							

b) Draw the graph of $y + x = 8$ on the grid opposite.

Q7 A rough way of changing temperatures from Celsius to Fahrenheit is to 'double it and add 30' ($y = 2x + 30$).

a) Fill in the table below to change the temperatures in Celsius (x) to Fahrenheit (y).

x	0	5	10	15	20	25	30
y	30				70		

b) Use your table of values to draw a temperature conversion graph. Draw and label the axes with x from 0 to 30, and y from 0 to 100.

Straight-Line Graphs

The gradient of a line is just a measure of the slope.
The box below tells you everything you need to know about it...

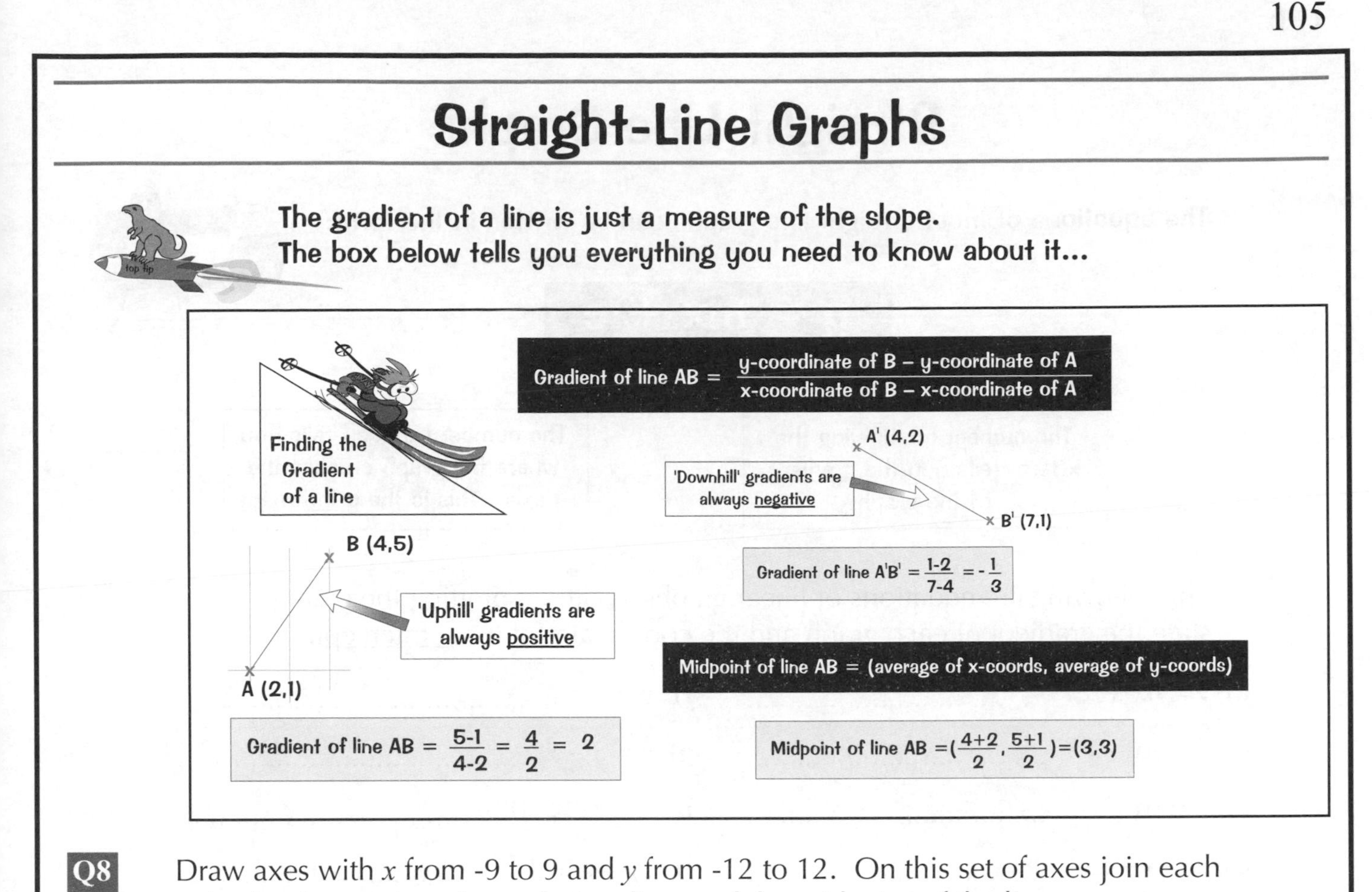

Q8 Draw axes with x from -9 to 9 and y from -12 to 12. On this set of axes join each pair of points and work out the gradient and the midpoint of the line.

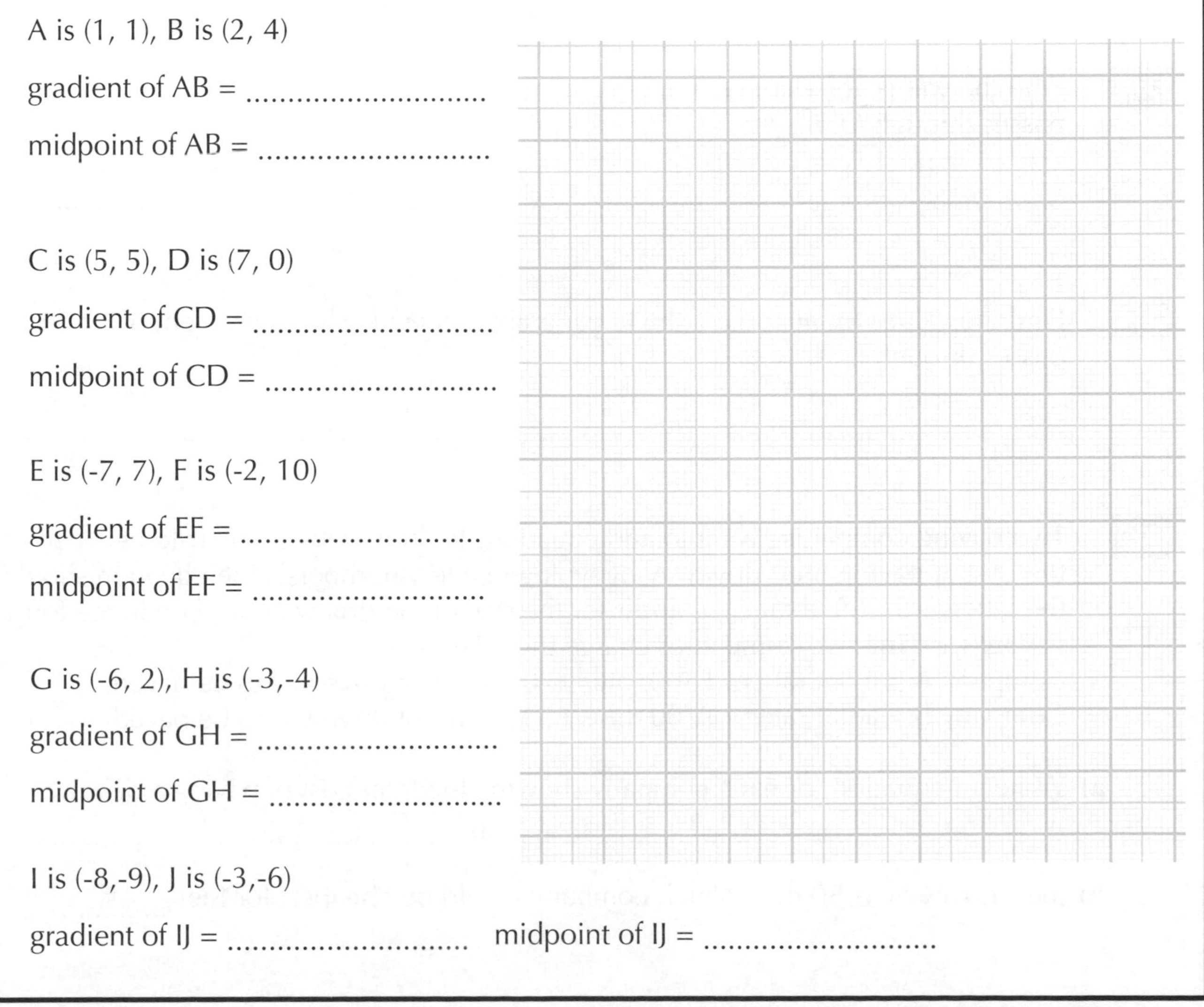

A is (1, 1), B is (2, 4)

gradient of AB = ...

midpoint of AB = ...

C is (5, 5), D is (7, 0)

gradient of CD = ...

midpoint of CD = ...

E is (-7, 7), F is (-2, 10)

gradient of EF = ...

midpoint of EF = ...

G is (-6, 2), H is (-3,-4)

gradient of GH = ...

midpoint of GH = ...

I is (-8,-9), J is (-3,-6)

gradient of IJ = ... midpoint of IJ = ...

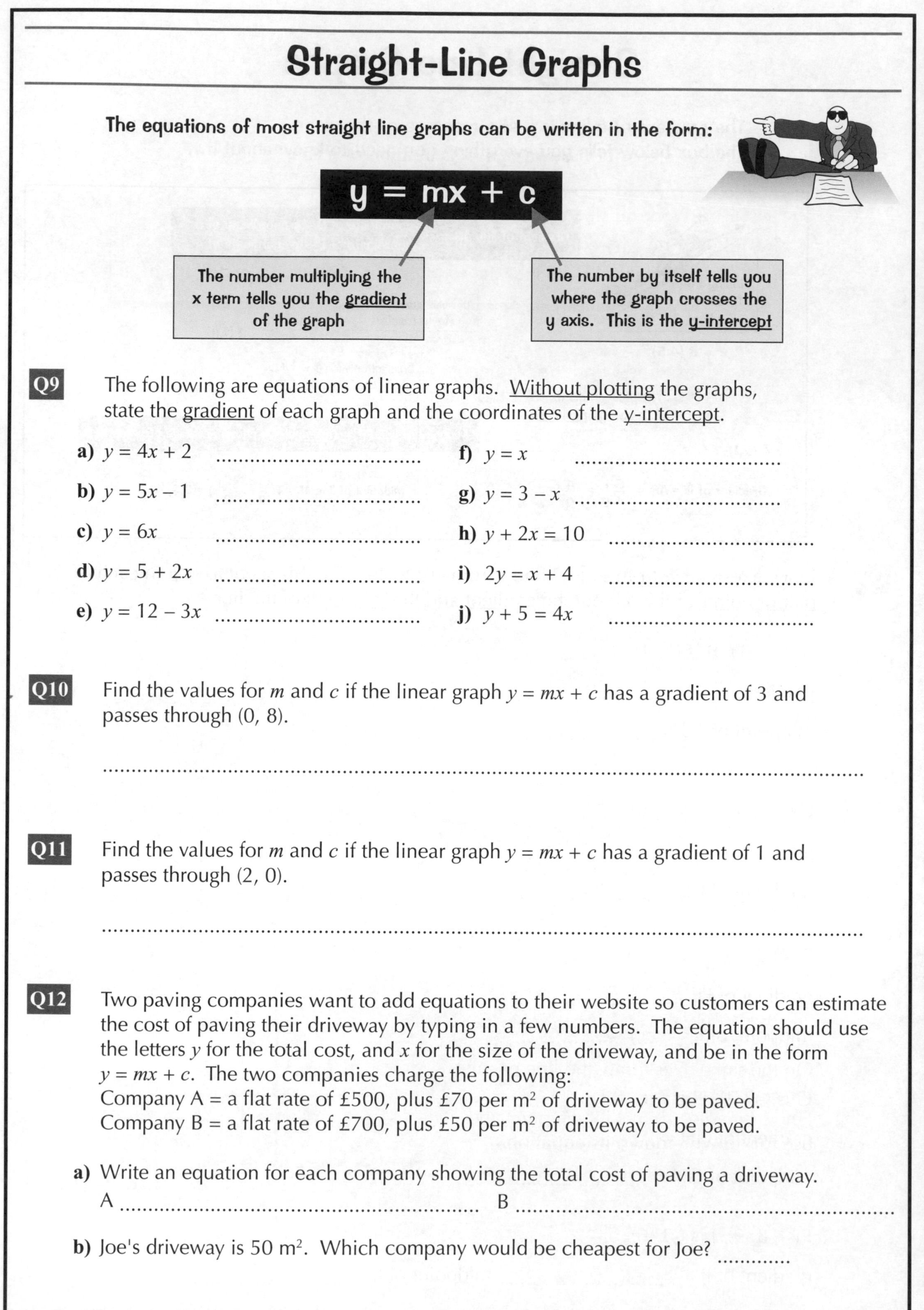

Straight-Line Graphs

The equations of most straight line graphs can be written in the form:

$$y = mx + c$$

The number multiplying the x term tells you the gradient of the graph

The number by itself tells you where the graph crosses the y axis. This is the y-intercept

Q9 The following are equations of linear graphs. Without plotting the graphs, state the gradient of each graph and the coordinates of the y-intercept.

a) $y = 4x + 2$ ....

b) $y = 5x - 1$ ....

c) $y = 6x$ ....

d) $y = 5 + 2x$ ....

e) $y = 12 - 3x$ ....

f) $y = x$ ....

g) $y = 3 - x$ ....

h) $y + 2x = 10$ ....

i) $2y = x + 4$ ....

j) $y + 5 = 4x$ ....

Q10 Find the values for m and c if the linear graph $y = mx + c$ has a gradient of 3 and passes through (0, 8).

....

Q11 Find the values for m and c if the linear graph $y = mx + c$ has a gradient of 1 and passes through (2, 0).

....

Q12 Two paving companies want to add equations to their website so customers can estimate the cost of paving their driveway by typing in a few numbers. The equation should use the letters y for the total cost, and x for the size of the driveway, and be in the form $y = mx + c$. The two companies charge the following:
Company A = a flat rate of £500, plus £70 per m^2 of driveway to be paved.
Company B = a flat rate of £700, plus £50 per m^2 of driveway to be paved.

a) Write an equation for each company showing the total cost of paving a driveway.
A .... B ....

b) Joe's driveway is 50 m^2. Which company would be cheapest for Joe? ....

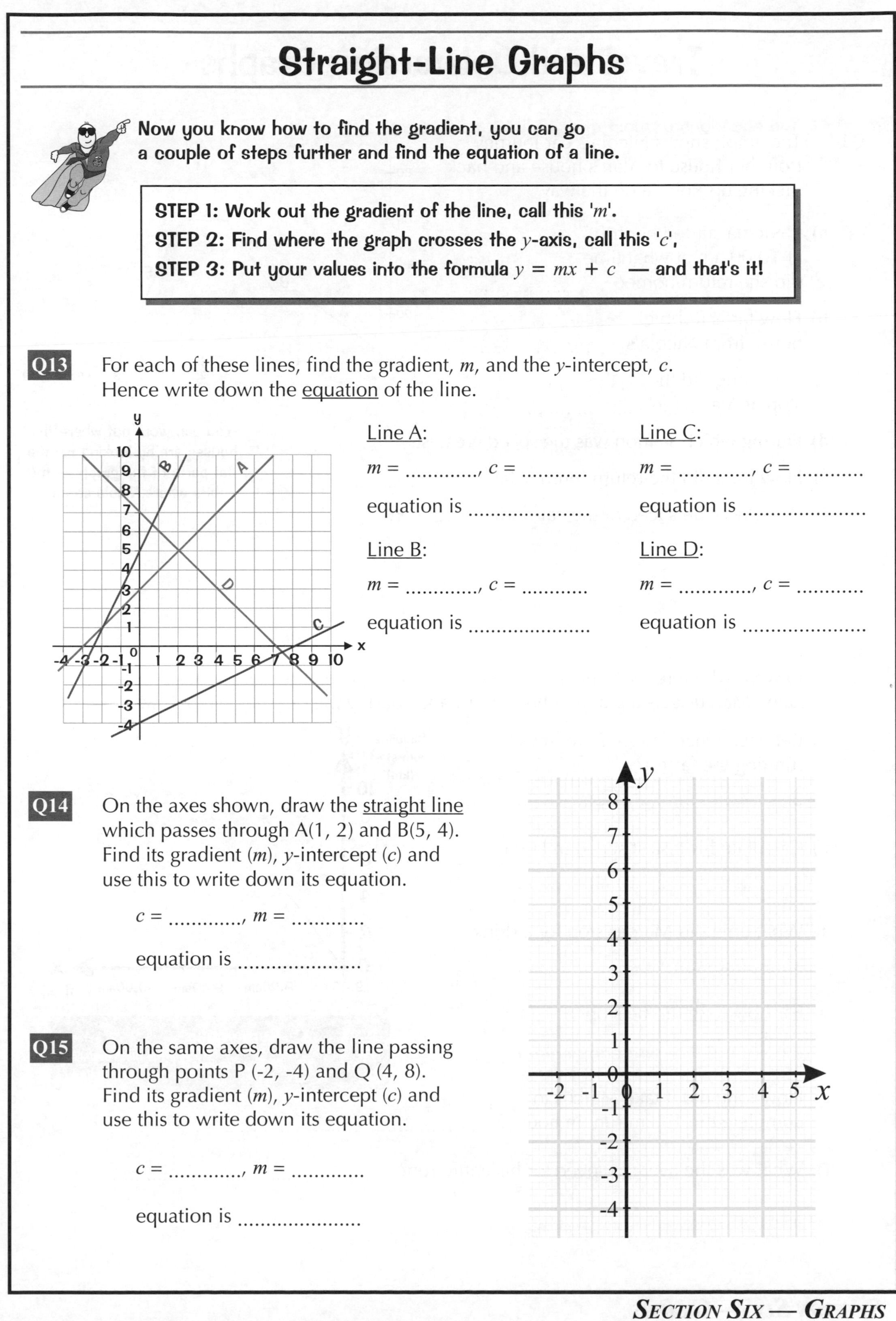

Straight-Line Graphs

Now you know how to find the gradient, you can go a couple of steps further and find the equation of a line.

STEP 1: Work out the gradient of the line, call this 'm'.
STEP 2: Find where the graph crosses the y-axis, call this 'c',
STEP 3: Put your values into the formula $y = mx + c$ — and that's it!

Q13 For each of these lines, find the gradient, m, and the y-intercept, c. Hence write down the equation of the line.

Line A:

$m =$ ...., $c =$ ....

equation is ....

Line B:

$m =$ ...., $c =$ ....

equation is ....

Line C:

$m =$ ...., $c =$ ....

equation is ....

Line D:

$m =$ ...., $c =$ ....

equation is ....

Q14 On the axes shown, draw the straight line which passes through A(1, 2) and B(5, 4). Find its gradient (m), y-intercept (c) and use this to write down its equation.

$c =$ ...., $m =$ ....

equation is ....

Q15 On the same axes, draw the line passing through points P (-2, -4) and Q (4, 8). Find its gradient (m), y-intercept (c) and use this to write down its equation.

$c =$ ...., $m =$ ....

equation is ....

Travel and Conversion Graphs

Q1 The graph shows Nicola's car journey from her house to Alan's house and back, picking up Robbie on the way.

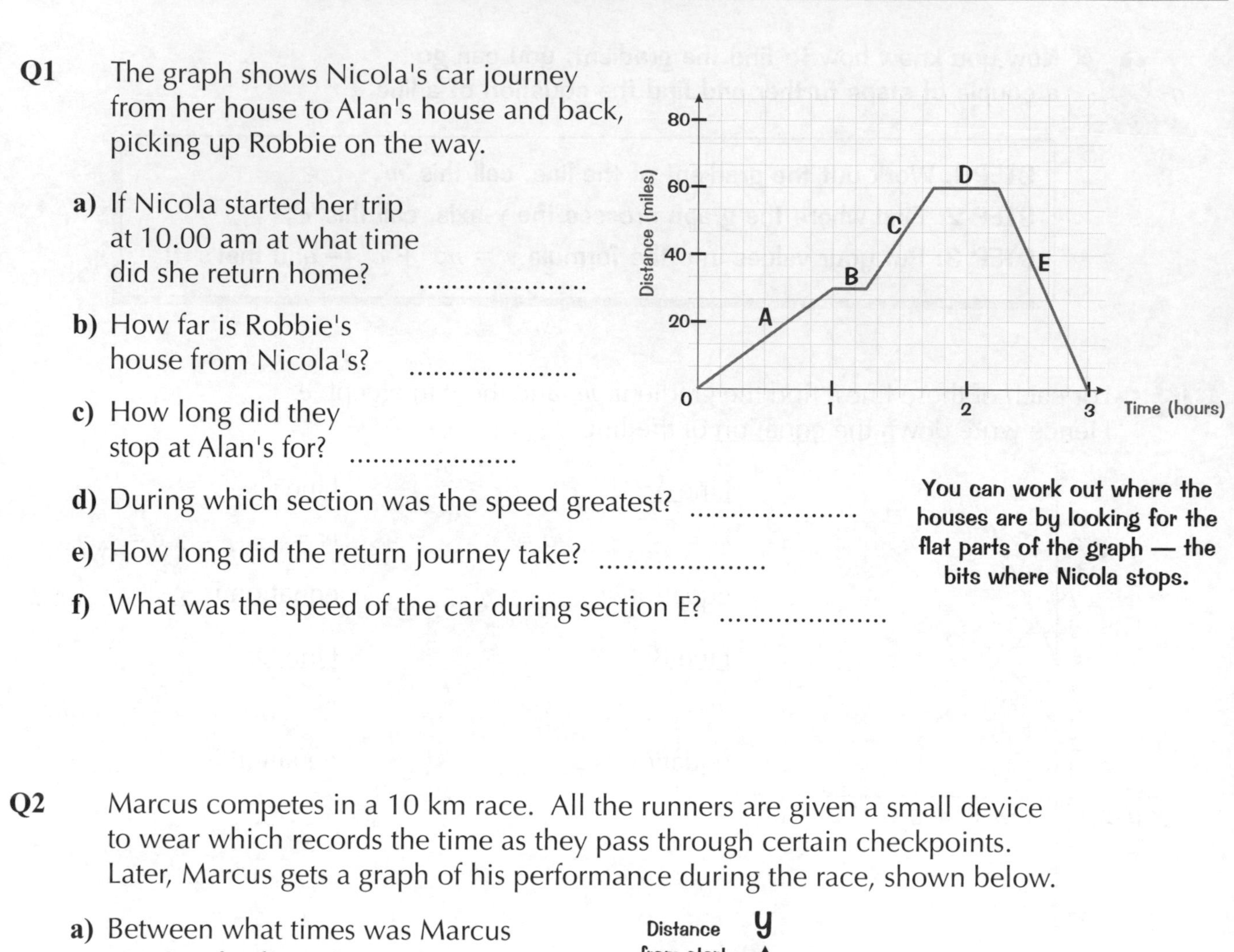

a) If Nicola started her trip at 10.00 am at what time did she return home? ....

b) How far is Robbie's house from Nicola's? ....

c) How long did they stop at Alan's for? ....

d) During which section was the speed greatest? ....

e) How long did the return journey take? ....

f) What was the speed of the car during section E? ....

You can work out where the houses are by looking for the flat parts of the graph — the bits where Nicola stops.

Q2 Marcus competes in a 10 km race. All the runners are given a small device to wear which records the time as they pass through certain checkpoints. Later, Marcus gets a graph of his performance during the race, shown below.

Distance from start (km) y

10
8
6
4
2
0

9.00am 9.20am 9.40am 10.00am x Time

a) Between what times was Marcus running the <u>fastest</u>?

....

b) Calculate his <u>fastest speed</u> in km/hr.

....

c) What time did Marcus <u>stop</u> for a drink?

....

d) For <u>how long</u> did he stop?

....

Remember:
Average speed = $\frac{\text{total dist. travelled}}{\text{total time taken}}$

e) How long did it take Marcus to <u>complete</u> the 10 km run (in hours)? ....

f) What was the <u>average speed</u> for his entire run? ....

Travel and Conversion Graphs

Q3 This graph can be used to convert the distance (miles) travelled in a taxi to the fare payable (£). How much will the fare be if you travel:

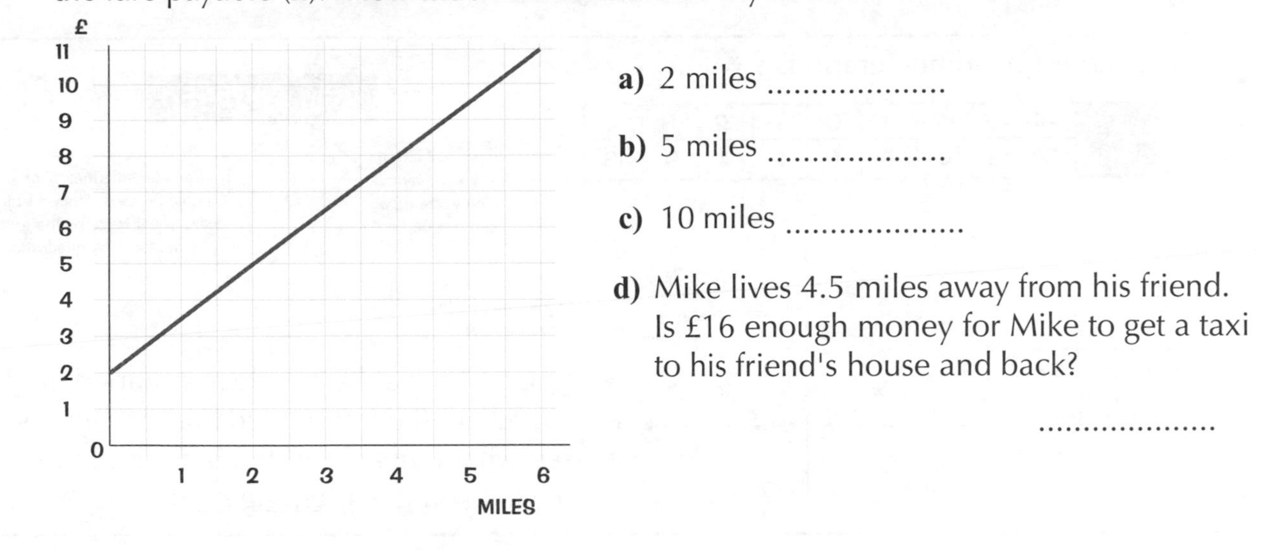

a) 2 miles ....

b) 5 miles ....

c) 10 miles ....

d) Mike lives 4.5 miles away from his friend. Is £16 enough money for Mike to get a taxi to his friend's house and back?

....

Q4 80 km is roughly equal to 50 miles. Use this information to draw a conversion graph on the grid. Use the graph to estimate the number of miles equal to:

a) 20 km ....

b) 70 km ....

c) 90 km ....

MILES: 0, 10, 20, 30, 40, 50, 60
KM: 20, 40, 60, 80, 100

Q5 How many km are equal to:

a) 40 miles ....

b) 10 miles ....

c) 30 miles ....

£: 0, 20, 40, 60, 80
litres: 10, 20, 30, 40, 50

Q6 Shelley fills up her car at a petrol station. Petrol costs her 150p per litre. Use this information to draw a conversion graph on the grid.

How much will it cost Shelley to fill her car up with 40 litres of petrol?

....

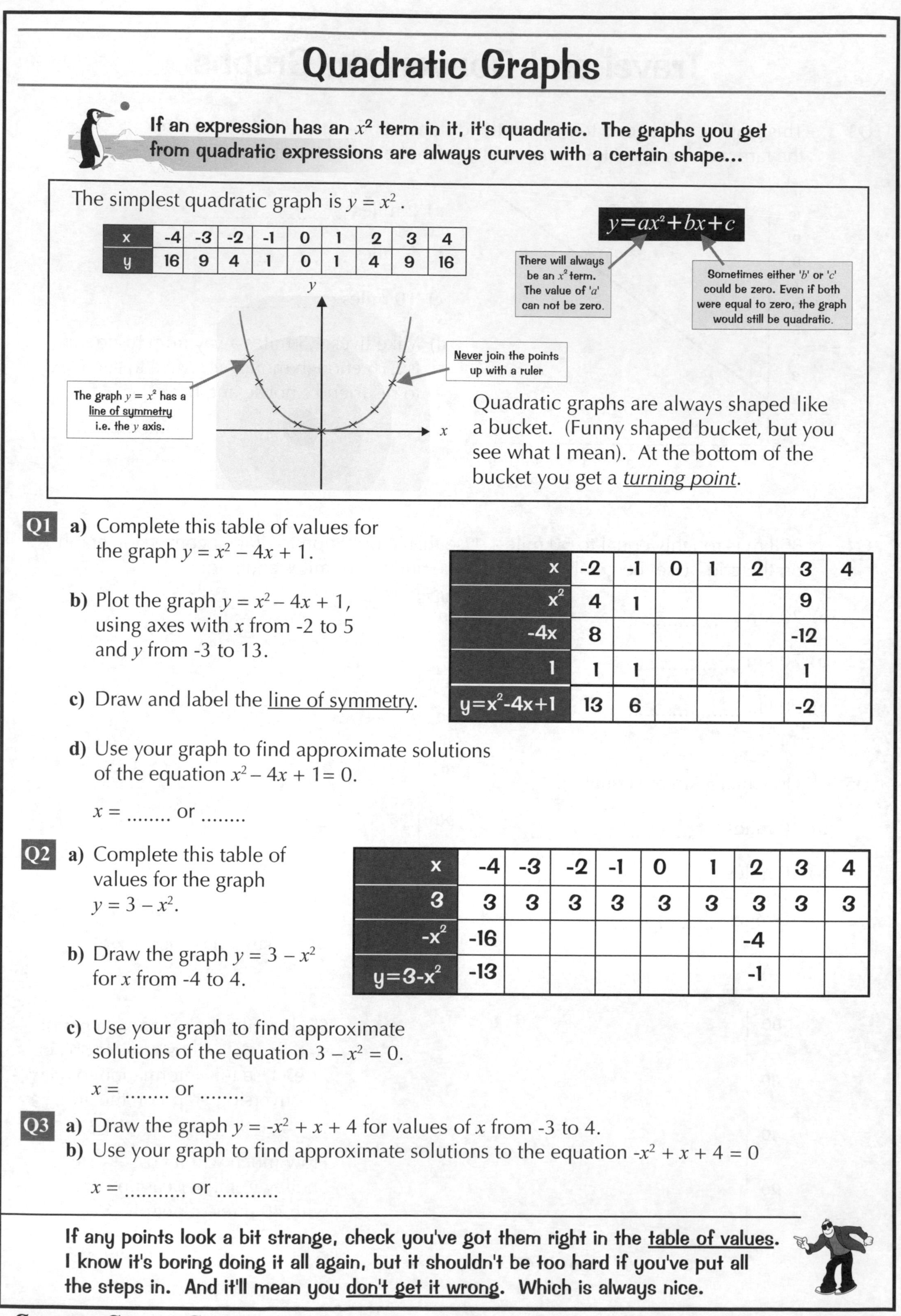

Quadratic Graphs

If an expression has an x^2 term in it, it's quadratic. The graphs you get from quadratic expressions are always curves with a certain shape...

The simplest quadratic graph is $y = x^2$.

x	-4	-3	-2	-1	0	1	2	3	4
y	16	9	4	1	0	1	4	9	16

Quadratic graphs are always shaped like a bucket. (Funny shaped bucket, but you see what I mean). At the bottom of the bucket you get a *turning point*.

Q1 **a)** Complete this table of values for the graph $y = x^2 - 4x + 1$.

x	-2	-1	0	1	2	3	4
x^2	4	1				9	
-4x	8					-12	
1	1	1				1	
$y=x^2-4x+1$	13	6				-2	

b) Plot the graph $y = x^2 - 4x + 1$, using axes with x from -2 to 5 and y from -3 to 13.

c) Draw and label the line of symmetry.

d) Use your graph to find approximate solutions of the equation $x^2 - 4x + 1 = 0$.

$x =$ .... or ....

Q2 **a)** Complete this table of values for the graph $y = 3 - x^2$.

x	-4	-3	-2	-1	0	1	2	3	4
3	3	3	3	3	3	3	3	3	3
$-x^2$	-16						-4		
$y=3-x^2$	-13						-1		

b) Draw the graph $y = 3 - x^2$ for x from -4 to 4.

c) Use your graph to find approximate solutions of the equation $3 - x^2 = 0$.

$x =$ .... or ....

Q3 **a)** Draw the graph $y = -x^2 + x + 4$ for values of x from -3 to 4.

b) Use your graph to find approximate solutions to the equation $-x^2 + x + 4 = 0$

$x =$ .... or ....

If any points look a bit strange, check you've got them right in the table of values. I know it's boring doing it all again, but it shouldn't be too hard if you've put all the steps in. And it'll mean you don't get it wrong. Which is always nice.

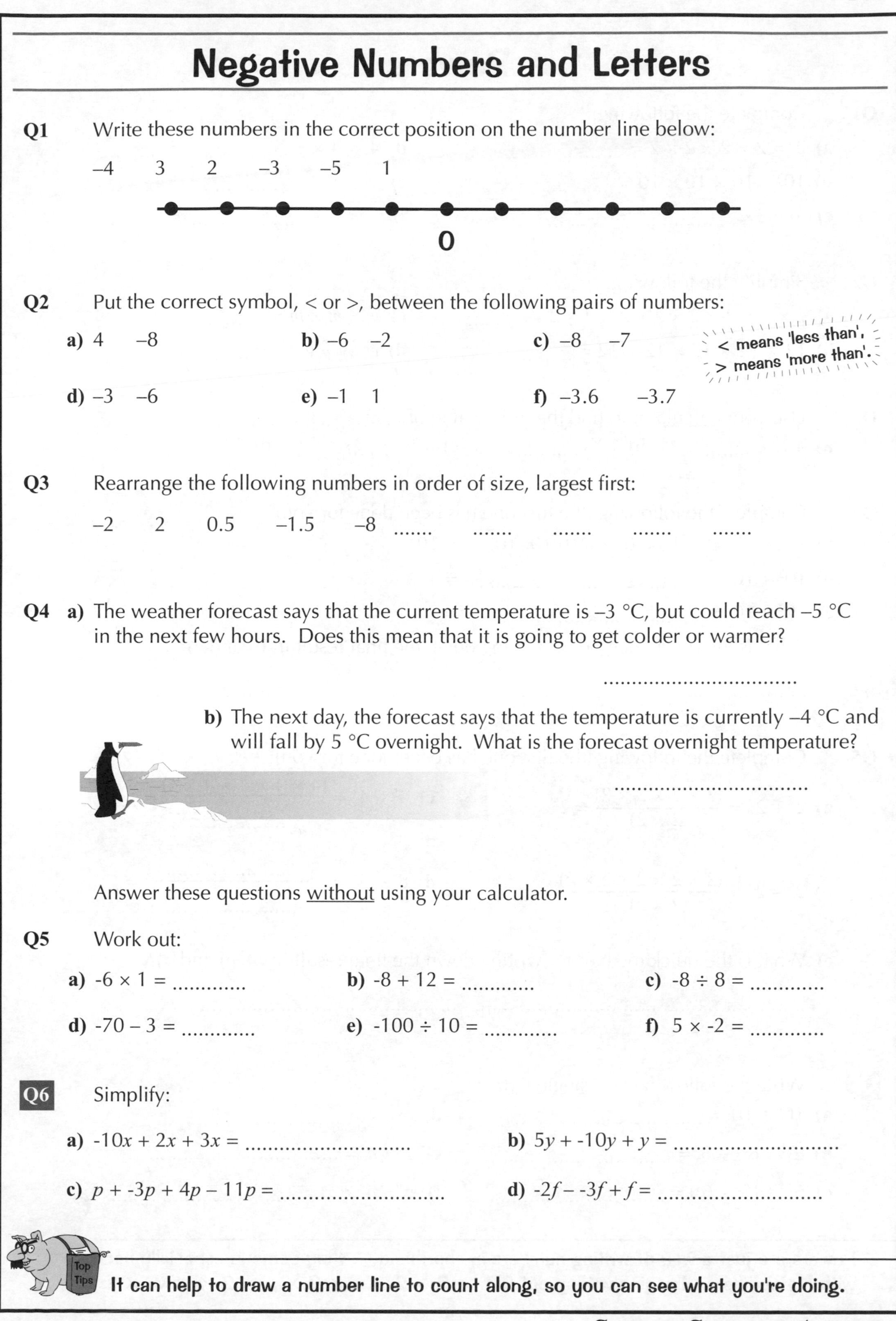

Negative Numbers and Letters

Q1 Write these numbers in the correct position on the number line below:

–4 3 2 –3 –5 1

0

Q2 Put the correct symbol, $<$ or $>$, between the following pairs of numbers:

a) 4 –8 **b)** –6 –2 **c)** –8 –7

d) –3 –6 **e)** –1 1 **f)** –3.6 –3.7

$<$ means 'less than', $>$ means 'more than'.

Q3 Rearrange the following numbers in order of size, largest first:

–2 2 0.5 –1.5 –8 ....

Q4 **a)** The weather forecast says that the current temperature is –3 °C, but could reach –5 °C in the next few hours. Does this mean that it is going to get colder or warmer?

....

b) The next day, the forecast says that the temperature is currently –4 °C and will fall by 5 °C overnight. What is the forecast overnight temperature?

....

Answer these questions <u>without</u> using your calculator.

Q5 Work out:

a) -6 × 1 = ... **b)** -8 + 12 = ... **c)** -8 ÷ 8 = ...

d) -70 – 3 = ... **e)** -100 ÷ 10 = ... **f)** 5 × -2 = ...

Q6 Simplify:

a) $-10x + 2x + 3x =$ .... **b)** $5y + -10y + y =$ ....

c) $p + -3p + 4p - 11p =$ .... **d)** $-2f - -3f + f =$ ....

Top Tips

It can help to draw a number line to count along, so you can see what you're doing.

Powers

Q1 Complete the following:

a) $2^4 = 2 \times 2 \times 2 \times 2 =$ ....

b) $10^3 = 10 \times 10 \times 10 =$ ....

c) $3^5 = 3\times$ .... = ...

d) $4^6 = 4 \times$ .... = ...

e) $1^9 = 1 \times$ .... = ...

f) $5^6 = 5 \times$ .... = ...

Q2 Simplify the following:

a) $2 \times 2 \times 2 \times 2 \times 2 \times 2 \times 2 \times 2 =$ ....

b) $12 \times 12 \times 12 \times 12 \times 12 =$ ....

c) $m \times m \times m =$ ....

d) $y \times y \times y \times y =$ ....

Q3 Use your calculator to find the exact value of

a) 4^3 .... **b)** 10^4 .... **c)** 12^5 .... **d)** 13^3 ....

Q4 Complete the following (the first one has been done for you):

a) $10^2 \times 10^3 = (10\times10) \times (10\times10\times10) = 10^5$

b) $10^3 \times 10^4 =$ .... =

c) $10^4 \times 10^2 =$ .... =

d) What is the quick method for writing down the final result in **b)** and **c)**?

....

Q5 Complete the following (the first one has been done for you):

a) $2^4 \div 2^2 = \frac{(2 \times 2 \times 2 \times 2)}{(2 \times 2)} = 2^2$

b) $2^5 \div 2^2 = \frac{(2 \times 2 \times 2 \times 2 \times 2)}{(2 \times 2)} =$ ....

c) $4^5 \div 4^3 = \frac{(4 \times 4 \times 4 \times 4 \times 4)}{\text{.......................}} =$ ....

d) $8^5 \div 8^2 = \frac{\text{.......................}}{\text{.......................}} =$ ....

e) What is the quick method for writing down the final result in **b), c)** and **d)**?

....

Q6 Write the following as a single term:

a) $10^6 \div 10^4 =$ ....

b) $(8^2 \times 8^5) \div 8^3 =$ ....

c) $6^{10} \div (6^2 \times 6^3) =$ ....

d) $x^2 \times x^3 =$ ....

e) $a^5 \times a^4 =$ ....

f) $p^4 \times p^5 \times p^6 =$ ....

Powers are just a way of writing numbers in shorthand — they come in especially handy with big numbers. Imagine writing out 2^{138} — 2 × 2 ×... × 2 ×... × 2 × yawn × zzz...

Square Roots and Cube Roots

Square root just means "WHAT NUMBER TIMES ITSELF (i.e. 2×2) GIVES..."
The square roots of 64 are 8 and –8 because 8×8=64 and -8×-8=64.
Cube root means "WHAT NUMBER TIMES ITSELF TWICE (i.e. 2×2×2) GIVES ..."
The cube root of 27 is 3 because 3×3×3=27.
Square roots always have a + and – answer, cube roots only have 1 answer.

Tip

Q1 Use the $\sqrt{\ }$ button on your calculator to find the following positive square roots to the nearest whole number.

a) $\sqrt{60}$ = ...
b) $\sqrt{19}$ = ...
c) $\sqrt{34}$ = ...
d) $\sqrt{200}$ = ...
e) $\sqrt{520}$ = ...
f) $\sqrt{75}$ = ...
g) $\sqrt{750}$ = ...
h) $\sqrt{0.9}$ = ...
i) $\sqrt{170}$ = ...
j) $\sqrt{7220}$ = ...
k) $\sqrt{1\,000\,050}$ = ...
l) $\sqrt{27}$ = ...

Q2 Without using a calculator, write down both answers to each of the following:

a) $\sqrt{4}$ = ...
b) $\sqrt{16}$ = ...
c) $\sqrt{9}$ = ...
d) $\sqrt{49}$ = ...
e) $\sqrt{25}$ = ...
f) $\sqrt{100}$ = ...
g) $\sqrt{144}$ = ...
h) $\sqrt{64}$ = ...
i) $\sqrt{81}$ = ...

Q3 Use your calculator to find the following:

a) $\sqrt[3]{4096}$ = ...
b) $\sqrt[3]{1728}$ = ...
c) $\sqrt[3]{1331}$ = ...
d) $\sqrt[3]{1\,000\,000}$ = ...
e) $\sqrt[3]{1}$ = ...
f) $\sqrt[3]{0.125}$ = ...

Q4 Without using a calculator, find the value of the following:

a) $\sqrt[3]{64}$ = ...
b) $\sqrt[3]{27}$ = ...
c) $\sqrt[3]{1000}$ = ...
d) $\sqrt[3]{8}$ = ...

Q5 Nida is buying a small gift box online. She sees a cube box with volume of 125 cm^3. What is the length of each box edge?

....

Q6 A farmer is buying fencing to surround a square field of area 3600 m^2. What length of fencing does he need to buy?

....

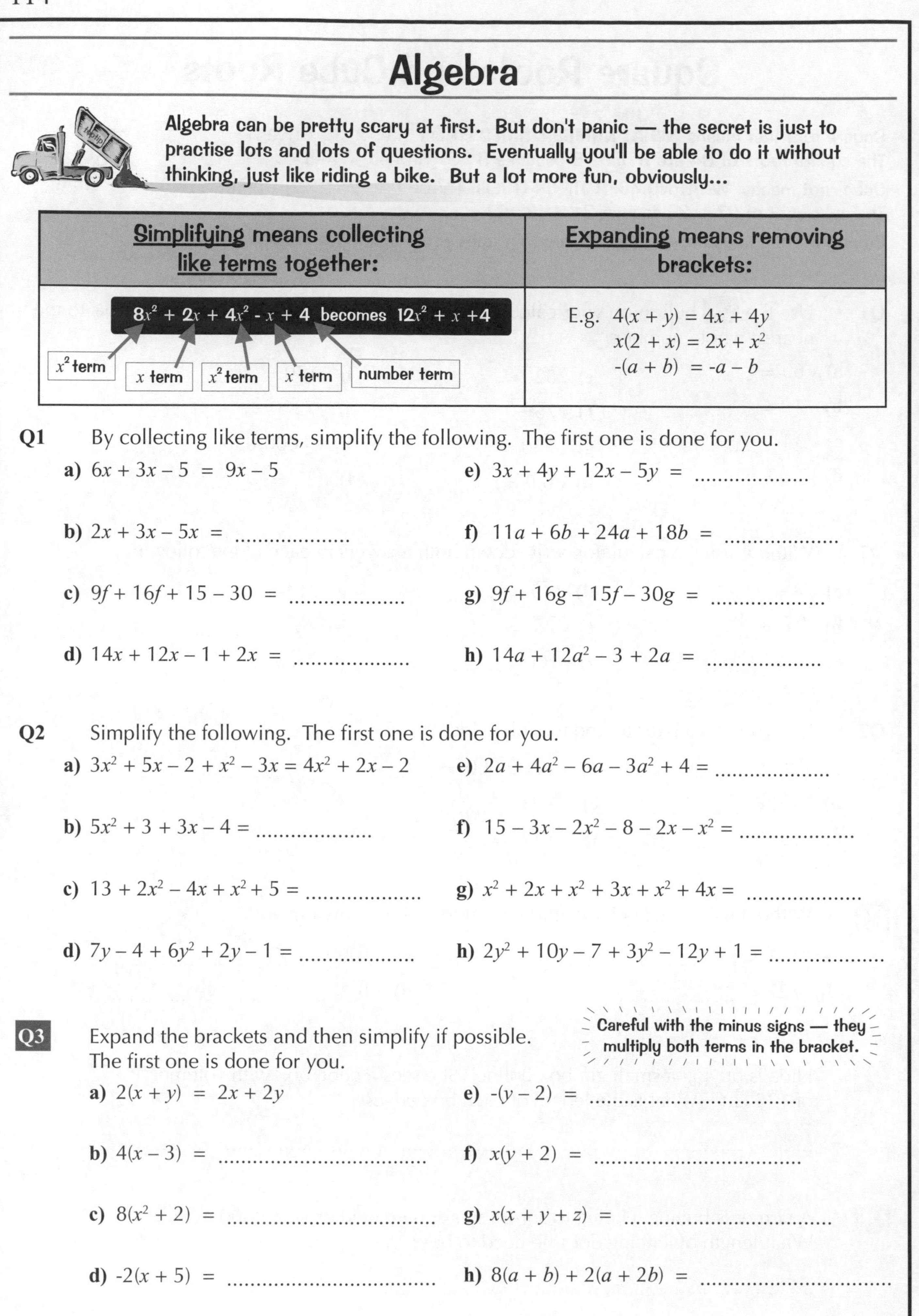

Algebra

Algebra can be pretty scary at first. But don't panic — the secret is just to practise lots and lots of questions. Eventually you'll be able to do it without thinking, just like riding a bike. But a lot more fun, obviously...

Simplifying means collecting **like terms** together:	**Expanding** means removing brackets:
$8x^2 + 2x + 4x^2 - x + 4$ becomes $12x^2 + x + 4$ (x^2 term, x term, x^2 term, x term, number term)	E.g. $4(x + y) = 4x + 4y$ $x(2 + x) = 2x + x^2$ $-(a + b) = -a - b$

Q1 By collecting like terms, simplify the following. The first one is done for you.

a) $6x + 3x - 5 = 9x - 5$

b) $2x + 3x - 5x =$ ....

c) $9f + 16f + 15 - 30 =$ ....

d) $14x + 12x - 1 + 2x =$ ....

e) $3x + 4y + 12x - 5y =$ ....

f) $11a + 6b + 24a + 18b =$ ....

g) $9f + 16g - 15f - 30g =$ ....

h) $14a + 12a^2 - 3 + 2a =$ ....

Q2 Simplify the following. The first one is done for you.

a) $3x^2 + 5x - 2 + x^2 - 3x = 4x^2 + 2x - 2$

b) $5x^2 + 3 + 3x - 4 =$ ....

c) $13 + 2x^2 - 4x + x^2 + 5 =$ ....

d) $7y - 4 + 6y^2 + 2y - 1 =$ ....

e) $2a + 4a^2 - 6a - 3a^2 + 4 =$ ....

f) $15 - 3x - 2x^2 - 8 - 2x - x^2 =$ ....

g) $x^2 + 2x + x^2 + 3x + x^2 + 4x =$ ....

h) $2y^2 + 10y - 7 + 3y^2 - 12y + 1 =$ ....

Q3 Expand the brackets and then simplify if possible. The first one is done for you.

Careful with the minus signs — they multiply both terms in the bracket.

a) $2(x + y) = 2x + 2y$

b) $4(x - 3) =$ ....

c) $8(x^2 + 2) =$ ....

d) $-2(x + 5) =$ ....

e) $-(y - 2) =$ ....

f) $x(y + 2) =$ ....

g) $x(x + y + z) =$ ....

h) $8(a + b) + 2(a + 2b) =$ ....

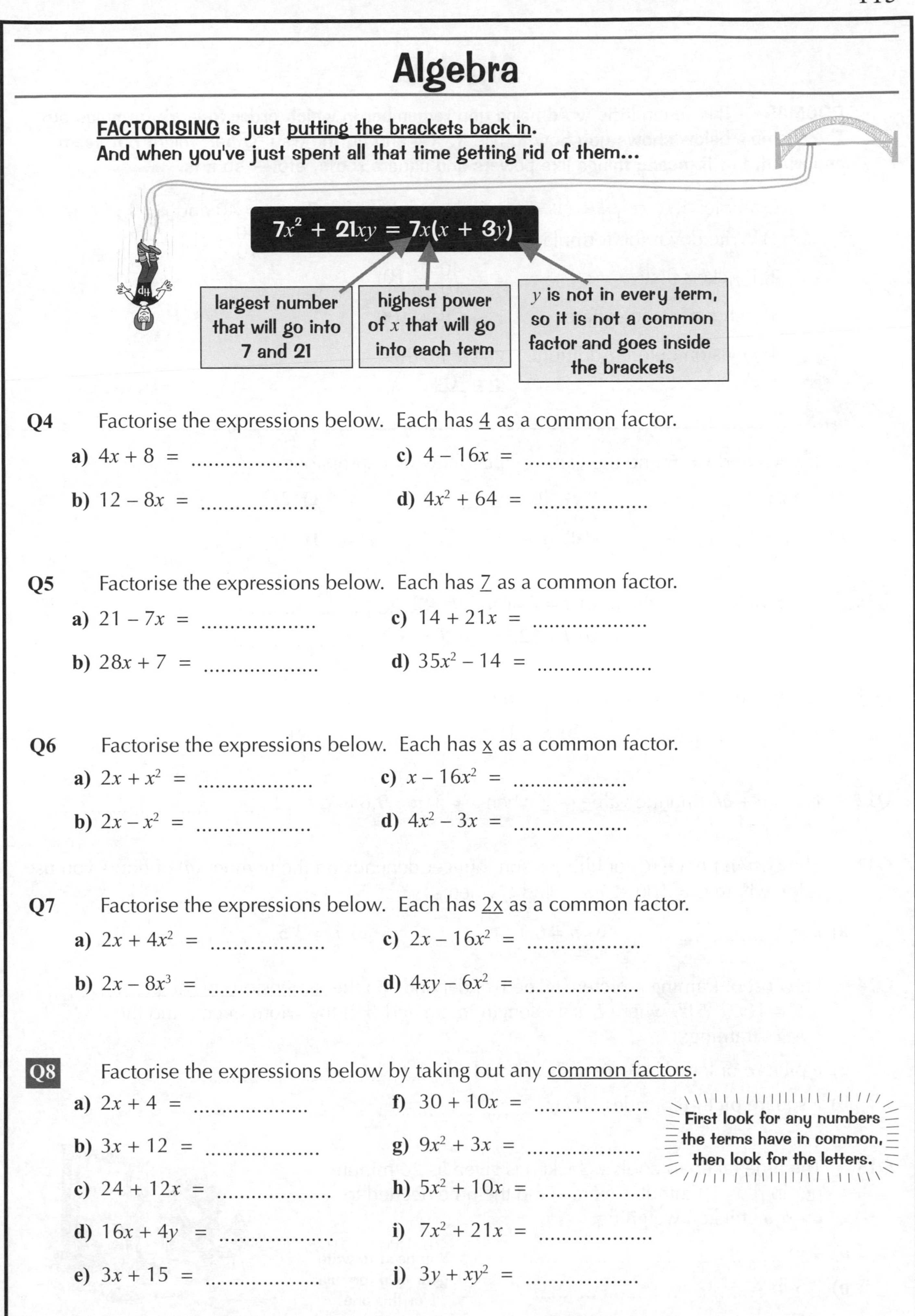

Algebra

FACTORISING is just putting the brackets back in.
And when you've just spent all that time getting rid of them...

$$7x^2 + 21xy = 7x(x + 3y)$$

Q4 Factorise the expressions below. Each has 4 as a common factor.

a) $4x + 8 =$ ....
c) $4 - 16x =$ ....
b) $12 - 8x =$ ....
d) $4x^2 + 64 =$ ....

Q5 Factorise the expressions below. Each has 7 as a common factor.

a) $21 - 7x =$ ....
c) $14 + 21x =$ ....
b) $28x + 7 =$ ....
d) $35x^2 - 14 =$ ....

Q6 Factorise the expressions below. Each has x as a common factor.

a) $2x + x^2 =$ ....
c) $x - 16x^2 =$ ....
b) $2x - x^2 =$ ....
d) $4x^2 - 3x =$ ....

Q7 Factorise the expressions below. Each has 2x as a common factor.

a) $2x + 4x^2 =$ ....
c) $2x - 16x^2 =$ ....
b) $2x - 8x^3 =$ ....
d) $4xy - 6x^2 =$ ....

Q8 Factorise the expressions below by taking out any common factors.

a) $2x + 4 =$ ....
b) $3x + 12 =$ ....
c) $24 + 12x =$ ....
d) $16x + 4y =$ ....
e) $3x + 15 =$ ....
f) $30 + 10x =$ ....
g) $9x^2 + 3x =$ ....
h) $5x^2 + 10x =$ ....
i) $7x^2 + 21x =$ ....
j) $3y + xy^2 =$ ....

First look for any numbers the terms have in common, then look for the letters.

Algebra

BODMAS — this funny little word helps you remember in which order to work formulas out. The example below shows you how to use it. Oh and by the way "Other" might not seem important, but it means things like powers and square roots, etc — so it is.

Example: if $z = \frac{x}{10} + (y - 3)^2$, find the value of z when $x = 40$ and $y = 13$.

1) Write down the formula with the numbers stuck in $z = \frac{40}{10} + (13 - 3)^2$,

2) Brackets first: $z = \frac{40}{10} + (10)^2$

3) Other next, so square: $z = \frac{40}{10} + 100$

4) Division before Addition: $z = 4 + 100$

$\underline{z = 104}$

Q9 If $x = 3$ and $y = 6$ find the value of the following expressions.

a) $x + 2y$ .... **c)** $4(x + y)$ .... **e)** $2x^2$ ....

b) $2x \div y$ .... **d)** $(y - x)^2$ .... **f)** $2y^2$ ....

Q10 If $V = lwh$, find V when, **a)** $l = 7$, $w = 5$, $h = 2$ ....

b) $l = 12$, $w = 8$, $h = 5$ ....

Q11 Using the formula $z = (x - 10)^2$, find the value of z when:

a) $x = 20$ .... **b)** $x = 15$ .... **c)** $x = -1$ ....

Q12 If $V = u + at$, find the value of V when $u = 8$, $a = 9.8$ and $t = 2$. ....

Q13 The cost in pence (C) of hiring a sun lounger depends on the number (n) of hours you use it for, where $C = 100 + 30n$. Find C when,

a) $n = 2$ .... **b)** $n = 6$ .... **c)** $n = 3.5$ ....

Q14 The cost of framing a picture, C pence, depends on the dimensions of the picture. If $C = 10L + 5W$, where L is the length in cm and W is the width in cm, find the cost of framing:

a) a picture of length 40 cm and width 24 cm ....

b) a square picture of sides 30 cm. ....

Q15 The time taken to cook a chicken is given as 20 minutes per lb plus 20 minutes extra. Find the time needed to cook a chicken weighing:

a) 4 lb ....

b) 7.5 lb ....

You need to write your own formula for this one.

Number Patterns and Sequences

Q1 Draw the next two pictures in each pattern.
How many match sticks are used in each picture?

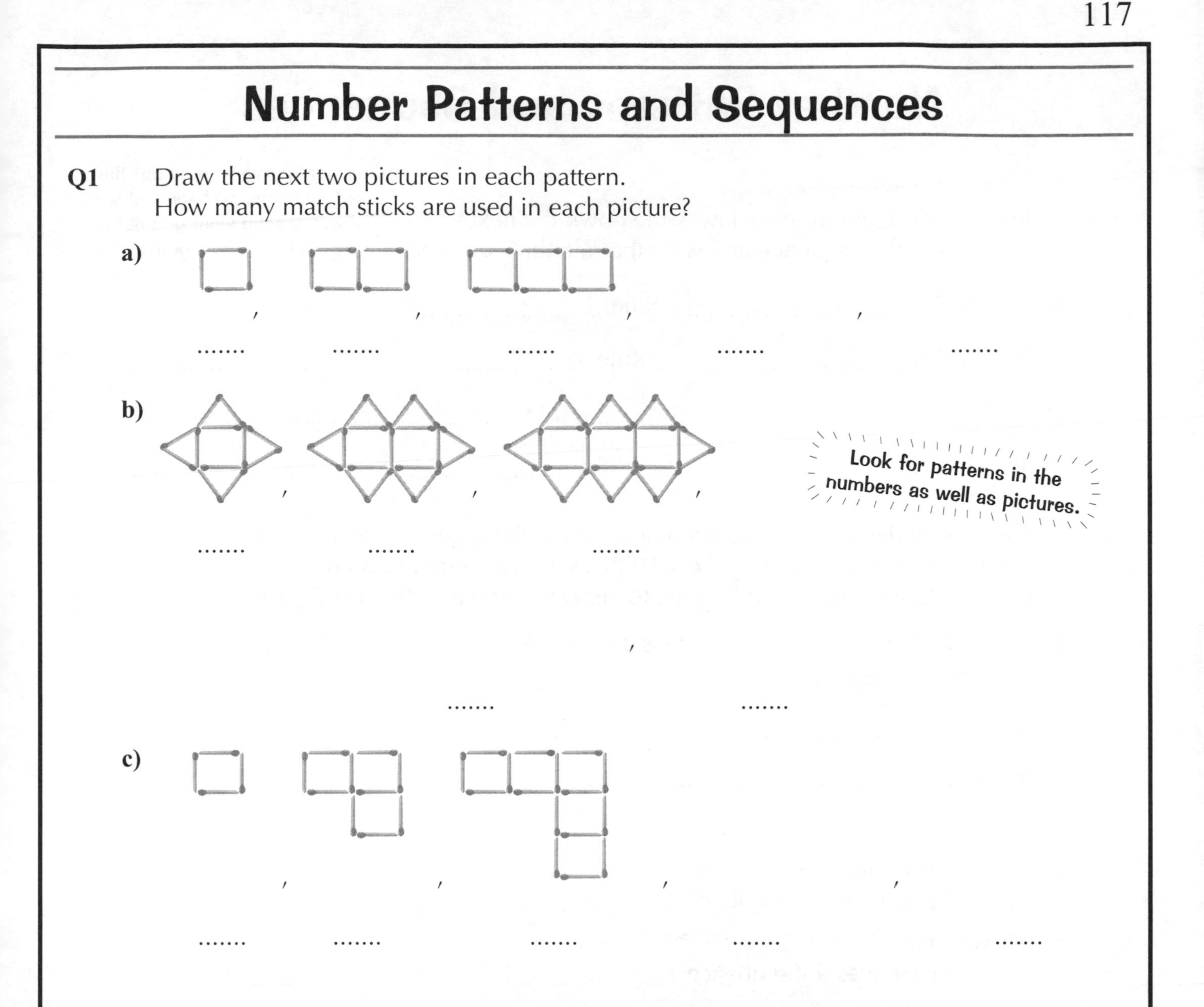

a) ...., ..., ..., ..., ...

b) ...., ..., ..., ..., ...

c) ...., ..., ..., ..., ...

Q2 Look for the pattern and then fill in the next three lines. Some of the answers are too big to fit on a calculator display so you must spot the pattern.

a)

$7 \times 6 = 42$

$67 \times 66 = 4422$

$667 \times 666 = 444\ 222$

$6667 \times 6666 =$ ....

$66\ 667 \times 66\ 666 =$ ....

$666\ 667 \times 666\ 666 =$ ....

b)

$1 \times 81 = 81$

$21 \times 81 = 1701$

$321 \times 81 = 26\ 001$

$4321 \times 81 = 350\ 001$

$54\ 321 \times 81 =$ ....

$654\ 321 \times 81 =$ ....

$7\ 654\ 321 \times 81 =$ ....

Q3 The first five terms of a sequence are 3, 7, 11, 15, 19...
Is 34 a term in the sequence? Explain your answer.

....

Number Patterns and Sequences

Q4 In each of the questions below, write down the next three numbers in the sequence and write the rule that you used.

Once you've worked out the next numbers, go back and write down exactly what you did — that will be the rule you're after.

a) 1, 3, 5, 7, ..., ..., ... Rule ....

b) 2, 4, 8, 16, ..., ..., ... Rule ....

c) 3, 30, 300, 3000, ..., ..., ... Rule ....

d) 3, 7, 11, 15, ..., ..., ... Rule ....

Q5 The letter ***n*** describes the position of a term in the sequence. For example, if ***n*** = 1, that's the 1st term... if ***n*** = 10 that's the 10th term and so on. In the following, use the rule given to generate (or make) the first 5 terms.

a) $3n + 1$ so if $n = 1$ the 1st term is <u>$(3 \times 1) + 1$</u> = <u>4</u>

$n = 2$ the 2nd term is .... = ...

$n = 3$ .... = ...

$n = 4$ .... = ...

$n = 5$ .... = ...

b) $5n - 2$, when $n = 1, 2, 3, 4$ and 5 produces the sequence ...., ..., ..., ..., ...

c) n^2, when $n = 1, 2, 3, 4$, and 5 produces the sequence ...., ..., ..., ..., ...

Q6 Write down an expression for the nth term of the following sequences:

Remember the formula — $dn + (a - d)$.

a) 2, 4, 6, 8, ...

....

b) 1, 3, 5, 7, ...

....

c) 5, 8, 11, 14, ...

....

Q7 Mike has opened a bank account to save for a holiday. He opened the account with £20 and puts £15 into the account at the end of each week. So at the end of the first week the balance of the account is £35. What will the balance of the account be after:

a) 3 weeks? .... **b)** 5 weeks? ....

c) Write down an expression that will let Mike work out how much he'll have in his account after any number of weeks he chooses.

....

d) What will the balance of the account be after 18 weeks?

....

Making Formulas from Words

It's no big mystery — algebra is just like normal sums, but with the odd letter or two stuck in for good measure.

Q1 Write the algebraic expression for these:

a) Three more than x ....

b) Seven less than y ....

c) Four multiplied by x ....

d) y multiplied by y ....

e) Ten divided by b ....

f) A number add five ....

Q2 Steven is 16 years old. How old will he be in:

a) 5 years? .... **b)** 10 years? .... **c)** x years? ....

Q3 Tickets for a football match cost £25 each. What is the cost for:

a) 2 tickets? ....

b) 6 tickets? ....

c) y tickets? ....

CGP Wanderers Football Club
Vs United Rovers FC
Comfy Seat
East stand lower bit
Row 20
Seat 104
£25.00

Q4 There are n books in a pile. Write an expression for the number of books in a pile that has:

a) 3 more books ....

b) 4 fewer books ....

c) Twice as many books ....

Q5 **a)** This square has sides of length 3 cm.

What is its perimeter? ....

What is its area? ....

3 cm

3 cm

b) This square has sides of length d cm.

What is its perimeter? ....

What is its area? ....

d cm

d cm

Q6 A cube has sides of length x cm. What is its volume? ....

Q7 The cost (C) of hiring a mountain bike is £10, plus £5 for each hour you use the bike (h). Write down a formula that can be used for working out the cost of hiring a bike.

....

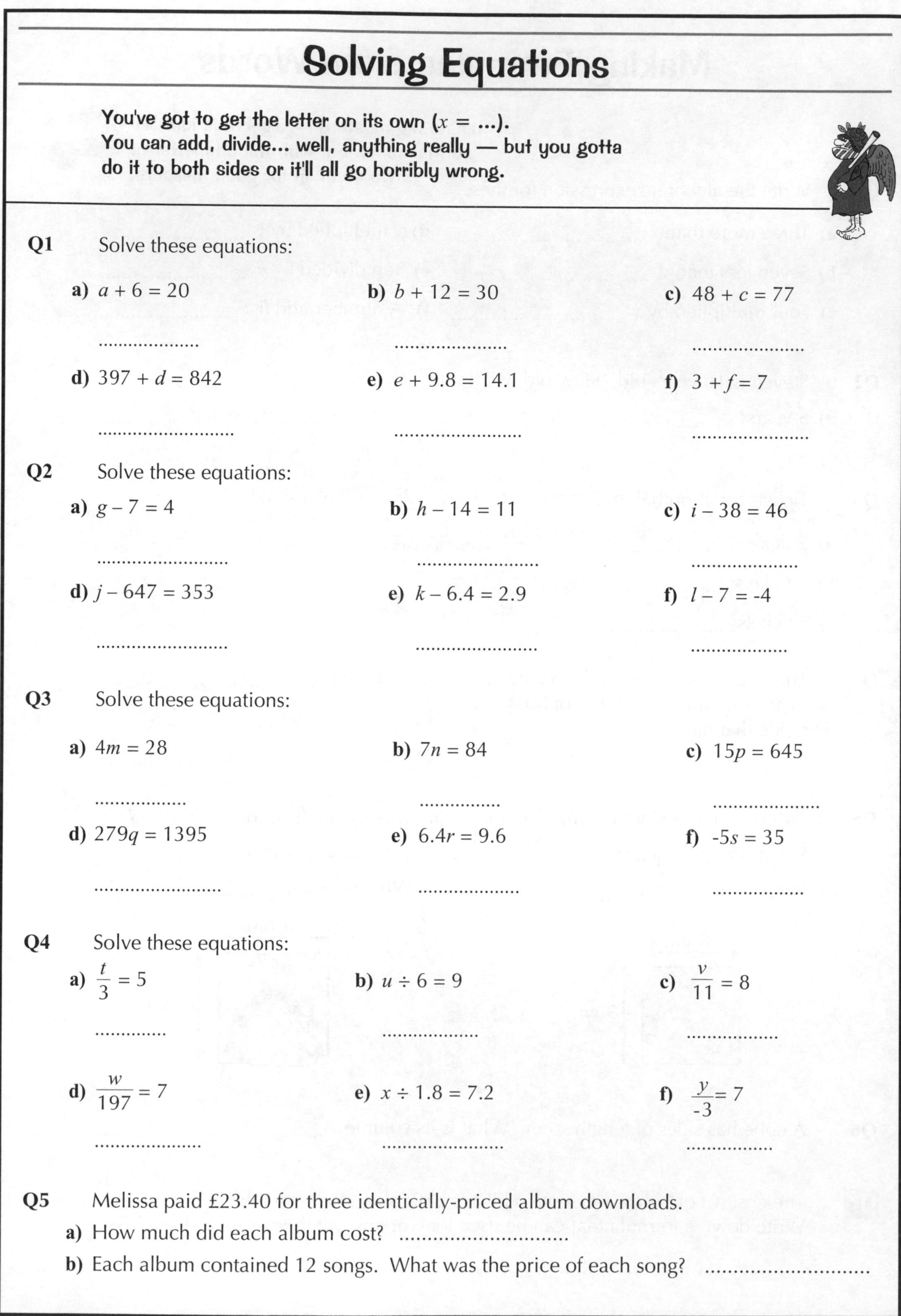

Solving Equations

**You've got to get the letter on its own (x = ...).
You can add, divide... well, anything really — but you gotta do it to both sides or it'll all go horribly wrong.**

Q1 Solve these equations:

a) $a + 6 = 20$ .... **b)** $b + 12 = 30$ .... **c)** $48 + c = 77$ ....

d) $397 + d = 842$ .... **e)** $e + 9.8 = 14.1$ .... **f)** $3 + f = 7$ ....

Q2 Solve these equations:

a) $g - 7 = 4$ .... **b)** $h - 14 = 11$ .... **c)** $i - 38 = 46$ ....

d) $j - 647 = 353$ .... **e)** $k - 6.4 = 2.9$ .... **f)** $l - 7 = -4$ ....

Q3 Solve these equations:

a) $4m = 28$ .... **b)** $7n = 84$ .... **c)** $15p = 645$ ....

d) $279q = 1395$ .... **e)** $6.4r = 9.6$ .... **f)** $-5s = 35$ ....

Q4 Solve these equations:

a) $\frac{t}{3} = 5$ .... **b)** $u \div 6 = 9$ .... **c)** $\frac{v}{11} = 8$ ....

d) $\frac{w}{197} = 7$ .... **e)** $x \div 1.8 = 7.2$ .... **f)** $\frac{y}{-3} = 7$ ....

Q5 Melissa paid £23.40 for three identically-priced album downloads.

a) How much did each album cost? ....

b) Each album contained 12 songs. What was the price of each song? ....

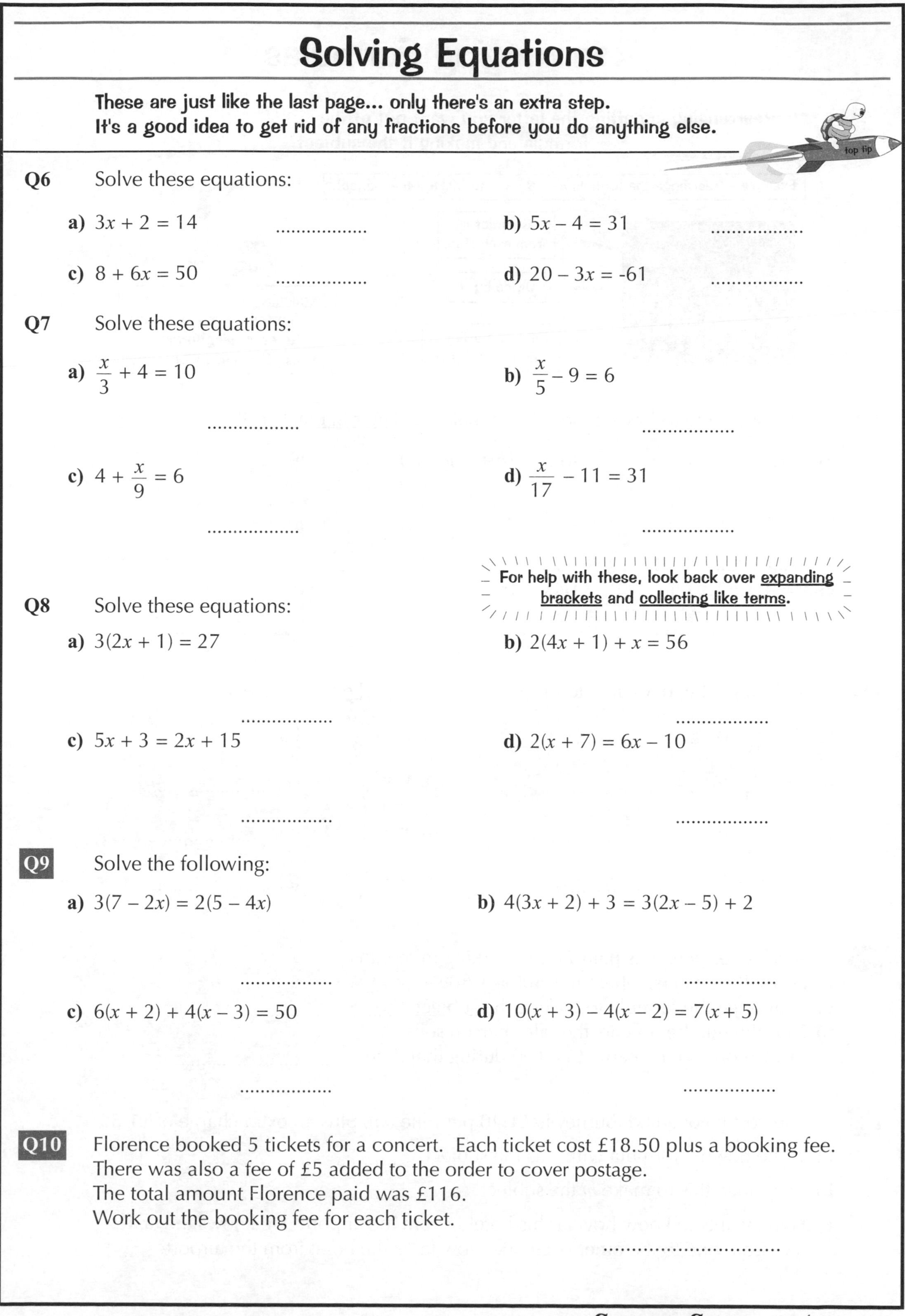

Solving Equations

These are just like the last page... only there's an extra step.
It's a good idea to get rid of any fractions before you do anything else.

Q6 Solve these equations:

a) $3x + 2 = 14$ ....

b) $5x - 4 = 31$ ....

c) $8 + 6x = 50$ ....

d) $20 - 3x = -61$ ....

Q7 Solve these equations:

a) $\frac{x}{3} + 4 = 10$ ....

b) $\frac{x}{5} - 9 = 6$ ....

c) $4 + \frac{x}{9} = 6$ ....

d) $\frac{x}{17} - 11 = 31$ ....

For help with these, look back over expanding brackets and collecting like terms.

Q8 Solve these equations:

a) $3(2x + 1) = 27$ ....

b) $2(4x + 1) + x = 56$ ....

c) $5x + 3 = 2x + 15$ ....

d) $2(x + 7) = 6x - 10$ ....

Q9 Solve the following:

a) $3(7 - 2x) = 2(5 - 4x)$ ....

b) $4(3x + 2) + 3 = 3(2x - 5) + 2$ ....

c) $6(x + 2) + 4(x - 3) = 50$ ....

d) $10(x + 3) - 4(x - 2) = 7(x + 5)$ ....

Q10 Florence booked 5 tickets for a concert. Each ticket cost £18.50 plus a booking fee.
There was also a fee of £5 added to the order to cover postage.
The total amount Florence paid was £116.
Work out the booking fee for each ticket.

....

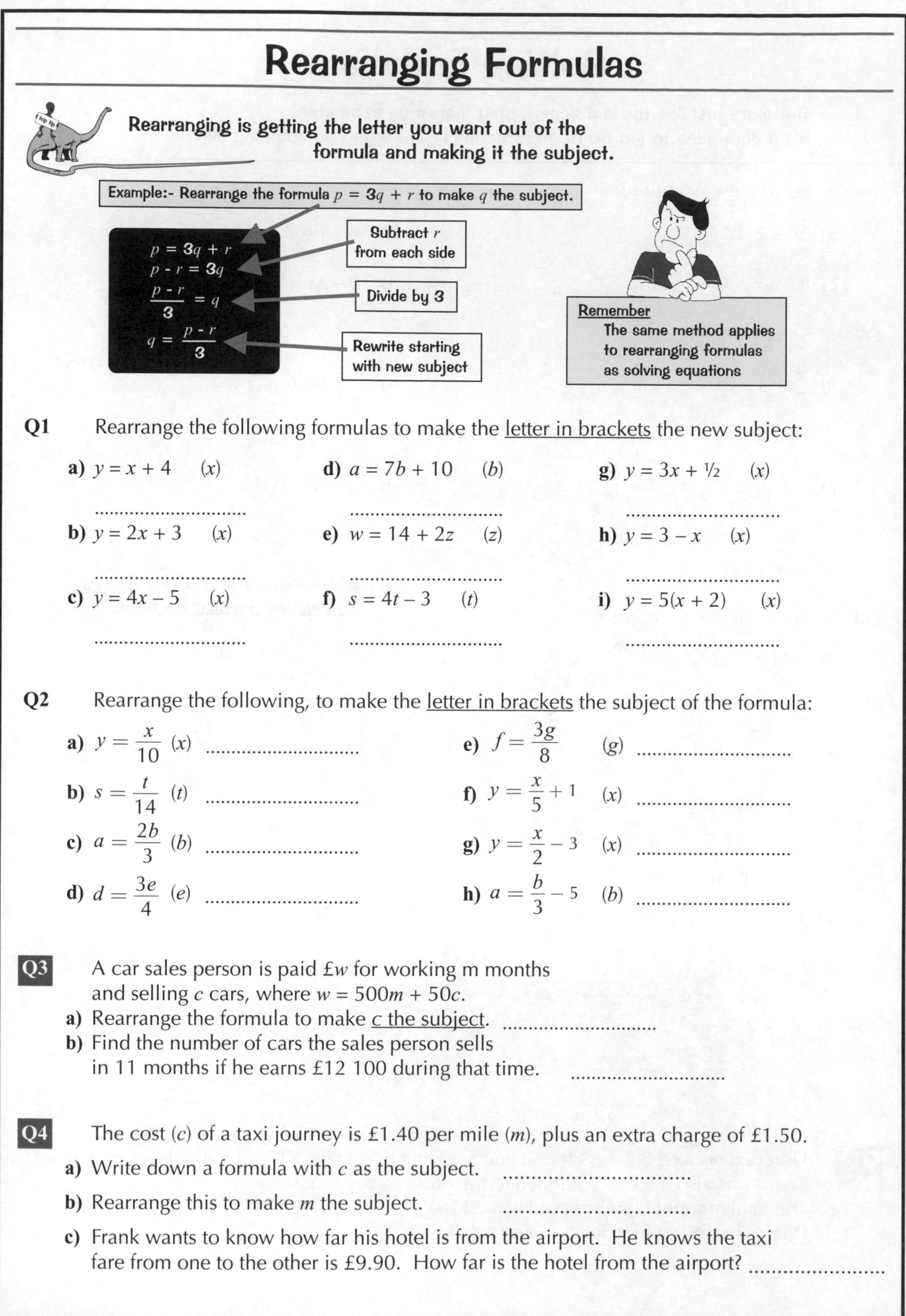

Rearranging Formulas

Rearranging is getting the letter you want out of the formula and making it the subject.

Example:- Rearrange the formula $p = 3q + r$ to make q the subject.

$p = 3q + r$

$p - r = 3q$

$\frac{p - r}{3} = q$

$q = \frac{p - r}{3}$

Subtract r from each side

Divide by 3

Rewrite starting with new subject

Remember
The same method applies to rearranging formulas as solving equations

Q1 Rearrange the following formulas to make the letter in brackets the new subject:

a) $y = x + 4$ (x) ....

b) $y = 2x + 3$ (x) ....

c) $y = 4x - 5$ (x) ....

d) $a = 7b + 10$ (b) ....

e) $w = 14 + 2z$ (z) ....

f) $s = 4t - 3$ (t) ....

g) $y = 3x + \frac{1}{2}$ (x) ....

h) $y = 3 - x$ (x) ....

i) $y = 5(x + 2)$ (x) ....

Q2 Rearrange the following, to make the letter in brackets the subject of the formula:

a) $y = \frac{x}{10}$ (x) ....

b) $s = \frac{t}{14}$ (t) ....

c) $a = \frac{2b}{3}$ (b) ....

d) $d = \frac{3e}{4}$ (e) ....

e) $f = \frac{3g}{8}$ (g) ....

f) $y = \frac{x}{5} + 1$ (x) ....

g) $y = \frac{x}{2} - 3$ (x) ....

h) $a = \frac{b}{3} - 5$ (b) ....

Q3 A car sales person is paid £w for working m months and selling c cars, where $w = 500m + 50c$.

a) Rearrange the formula to make c the subject. ....

b) Find the number of cars the sales person sells in 11 months if he earns £12 100 during that time. ....

Q4 The cost (c) of a taxi journey is £1.40 per mile (m), plus an extra charge of £1.50.

a) Write down a formula with c as the subject. ....

b) Rearrange this to make m the subject. ....

c) Frank wants to know how far his hotel is from the airport. He knows the taxi fare from one to the other is £9.90. How far is the hotel from the airport? ....

Trial and Improvement

Q1 Use the trial and improvement method to solve the equation $x^3 = 50$.
Give your answer to one decimal place. Two trials have been done for you.

Try $x = 3$ $x^3 = 27$ (too small)
Try $x = 4$ $x^3 = 64$ (too big)

....

Q2 Use the trial and improvement method to solve these equations.
Give your answers to one decimal place.

a) $x^2 + x = 80$

....

b) $x^3 - x = 100$

....

Show all the numbers you've tried, not just your final answer... or you'll be chucking away easy marks.

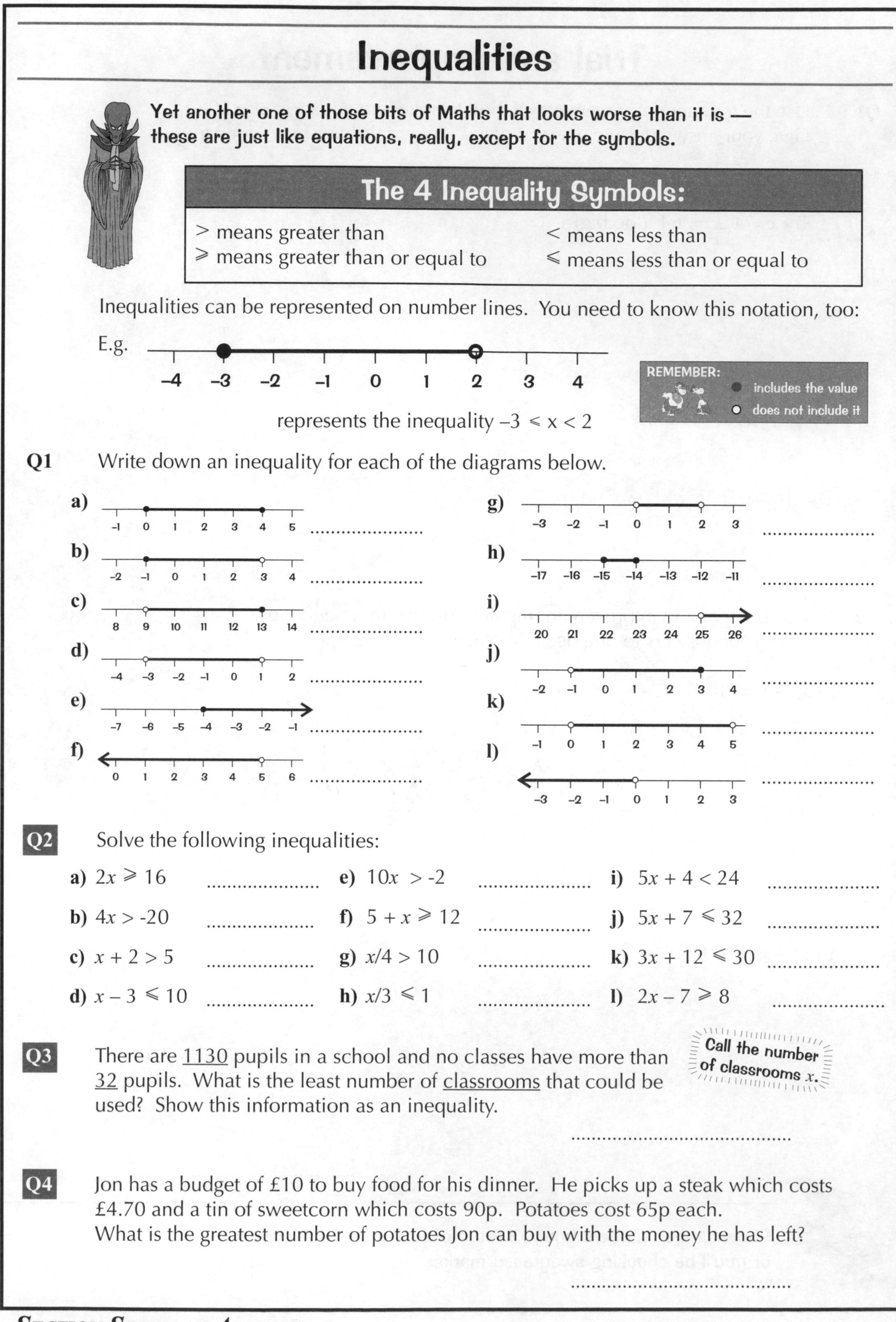

Inequalities

Yet another one of those bits of Maths that looks worse than it is — these are just like equations, really, except for the symbols.

The 4 Inequality Symbols:

$>$ means greater than	$<$ means less than
$\geqslant$ means greater than or equal to	$\leqslant$ means less than or equal to

Inequalities can be represented on number lines. You need to know this notation, too:

E.g.

represents the inequality $-3 \leqslant x < 2$

Q1 Write down an inequality for each of the diagrams below.

a) –1 0 1 2 3 4 5 ....

b) –2 –1 0 1 2 3 4 ....

c) 8 9 10 11 12 13 14 ....

d) –4 –3 –2 –1 0 1 2 ....

e) –7 –6 –5 –4 –3 –2 –1 ....

f) 0 1 2 3 4 5 6 ....

g) –3 –2 –1 0 1 2 3 ....

h) –17 –16 –15 –14 –13 –12 –11 ....

i) 20 21 22 23 24 25 26 ....

j) –2 –1 0 1 2 3 4 ....

k) –1 0 1 2 3 4 5 ....

l) –3 –2 –1 0 1 2 3 ....

Q2 Solve the following inequalities:

a) $2x \geqslant 16$ ....

b) $4x > -20$ ....

c) $x + 2 > 5$ ....

d) $x - 3 \leqslant 10$ ....

e) $10x > -2$ ....

f) $5 + x \geqslant 12$ ....

g) $x/4 > 10$ ....

h) $x/3 \leqslant 1$ ....

i) $5x + 4 < 24$ ....

j) $5x + 7 \leqslant 32$ ....

k) $3x + 12 \leqslant 30$ ....

l) $2x - 7 \geqslant 8$ ....

Q3 There are 1130 pupils in a school and no classes have more than 32 pupils. What is the least number of classrooms that could be used? Show this information as an inequality.

Call the number of classrooms x.

....

Q4 Jon has a budget of £10 to buy food for his dinner. He picks up a steak which costs £4.70 and a tin of sweetcorn which costs 90p. Potatoes cost 65p each. What is the greatest number of potatoes Jon can buy with the money he has left?

....

Travel and Conversion Graphs

Q3 This graph can be used to convert the distance (miles) travelled in a taxi to the fare payable (£). How much will the fare be if you travel:

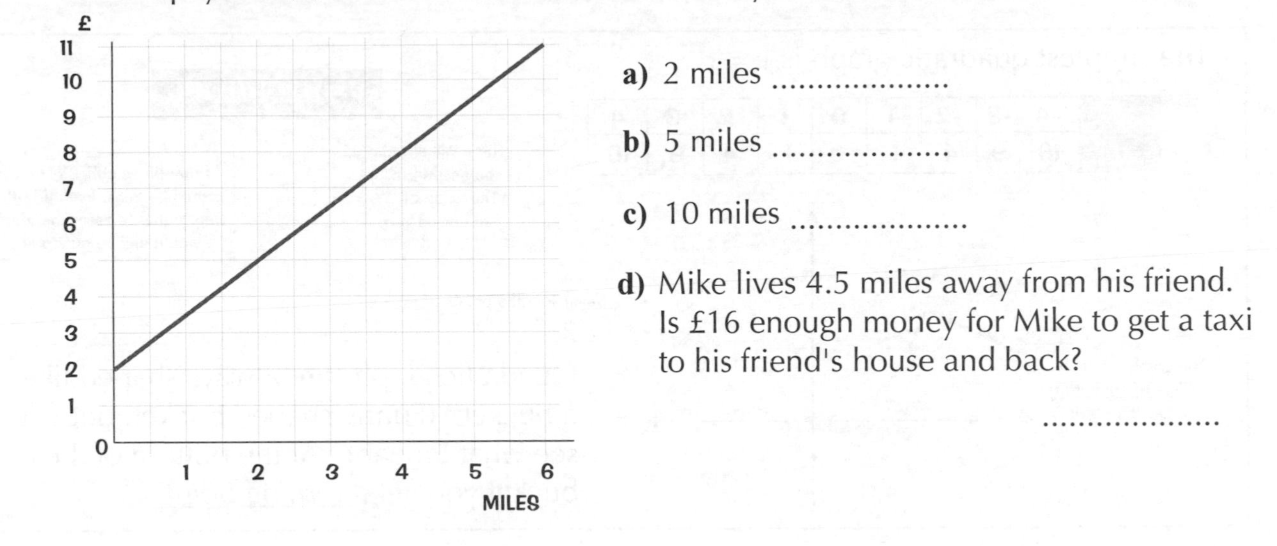

a) 2 miles ....

b) 5 miles ....

c) 10 miles ....

d) Mike lives 4.5 miles away from his friend. Is £16 enough money for Mike to get a taxi to his friend's house and back?

....

Q4 80 km is roughly equal to 50 miles. Use this information to draw a conversion graph on the grid. Use the graph to estimate the number of miles equal to:

a) 20 km ....

b) 70 km ....

c) 90 km ....

MILES
60
50
40
30
20
10
0
20 40 60 80 100 KM

Q5 How many km are equal to:

a) 40 miles ....

b) 10 miles ....

c) 30 miles ....

£
80
60
40
20
0
10 20 30 40 50 litres

Q6 Shelley fills up her car at a petrol station. Petrol costs her 150p per litre. Use this information to draw a conversion graph on the grid.

How much will it cost Shelley to fill her car up with 40 litres of petrol?

....

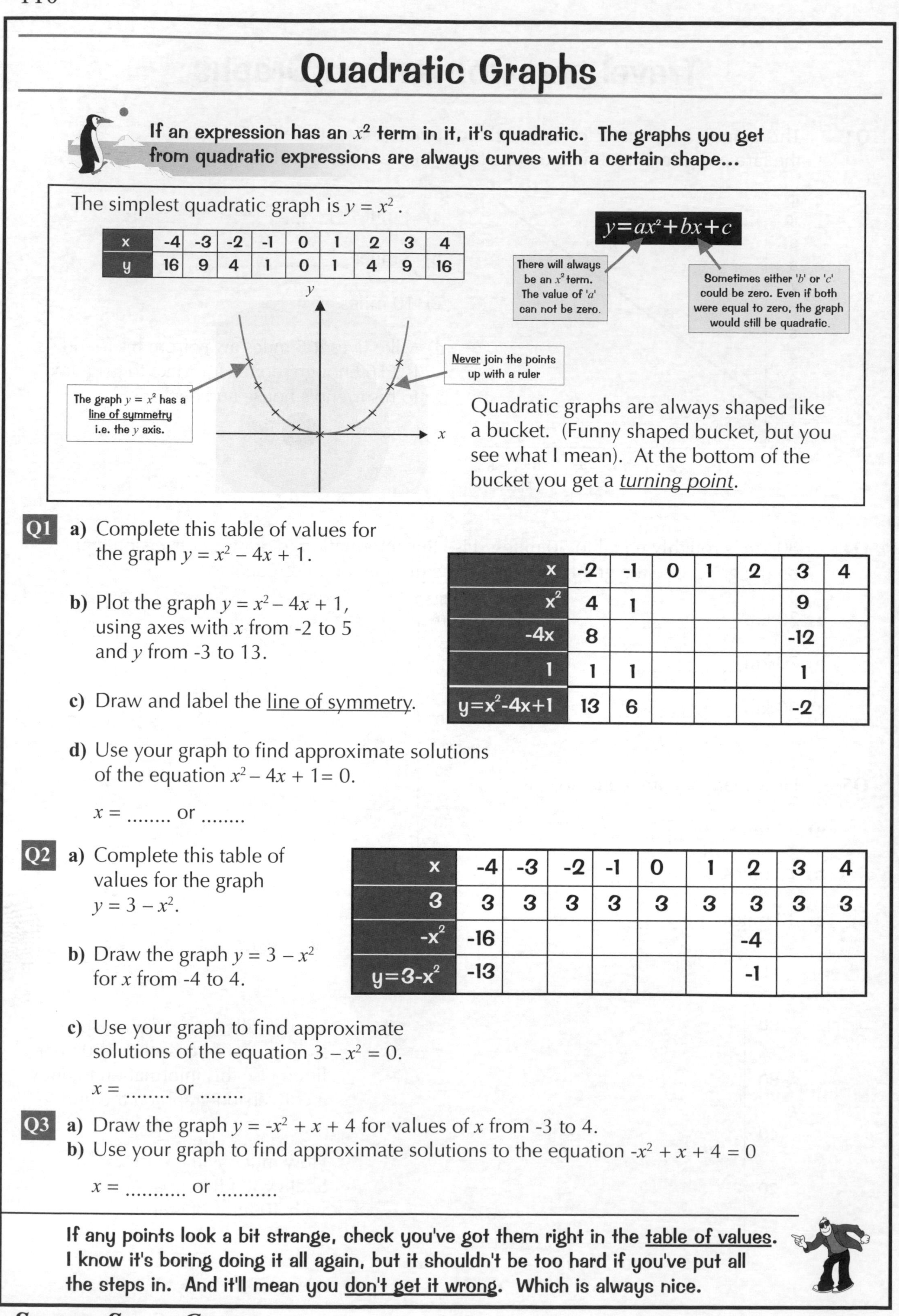

Quadratic Graphs

If an expression has an x^2 term in it, it's quadratic. The graphs you get from quadratic expressions are always curves with a certain shape...

The simplest quadratic graph is $y = x^2$.

x	-4	-3	-2	-1	0	1	2	3	4
y	16	9	4	1	0	1	4	9	16

$y=ax^2+bx+c$

There will always be an x^2 term. The value of 'a' can not be zero.

Sometimes either 'b' or 'c' could be zero. Even if both were equal to zero, the graph would still be quadratic.

Quadratic graphs are always shaped like a bucket. (Funny shaped bucket, but you see what I mean). At the bottom of the bucket you get a turning point.

Q1 **a)** Complete this table of values for the graph $y = x^2 - 4x + 1$.

x	-2	-1	0	1	2	3	4
x^2	4	1				9	
-4x	8					-12	
1	1	1				1	
$y=x^2-4x+1$	13	6				-2	

b) Plot the graph $y = x^2 - 4x + 1$, using axes with x from -2 to 5 and y from -3 to 13.

c) Draw and label the line of symmetry.

d) Use your graph to find approximate solutions of the equation $x^2 - 4x + 1 = 0$.

$x =$ .... or ....

Q2 **a)** Complete this table of values for the graph $y = 3 - x^2$.

x	-4	-3	-2	-1	0	1	2	3	4
3	3	3	3	3	3	3	3	3	3
$-x^2$	-16						-4		
$y=3-x^2$	-13						-1		

b) Draw the graph $y = 3 - x^2$ for x from -4 to 4.

c) Use your graph to find approximate solutions of the equation $3 - x^2 = 0$.

$x =$ .... or ....

Q3 **a)** Draw the graph $y = -x^2 + x + 4$ for values of x from -3 to 4.
b) Use your graph to find approximate solutions to the equation $-x^2 + x + 4 = 0$

$x =$ .... or ....

If any points look a bit strange, check you've got them right in the table of values. I know it's boring doing it all again, but it shouldn't be too hard if you've put all the steps in. And it'll mean you don't get it wrong. Which is always nice.

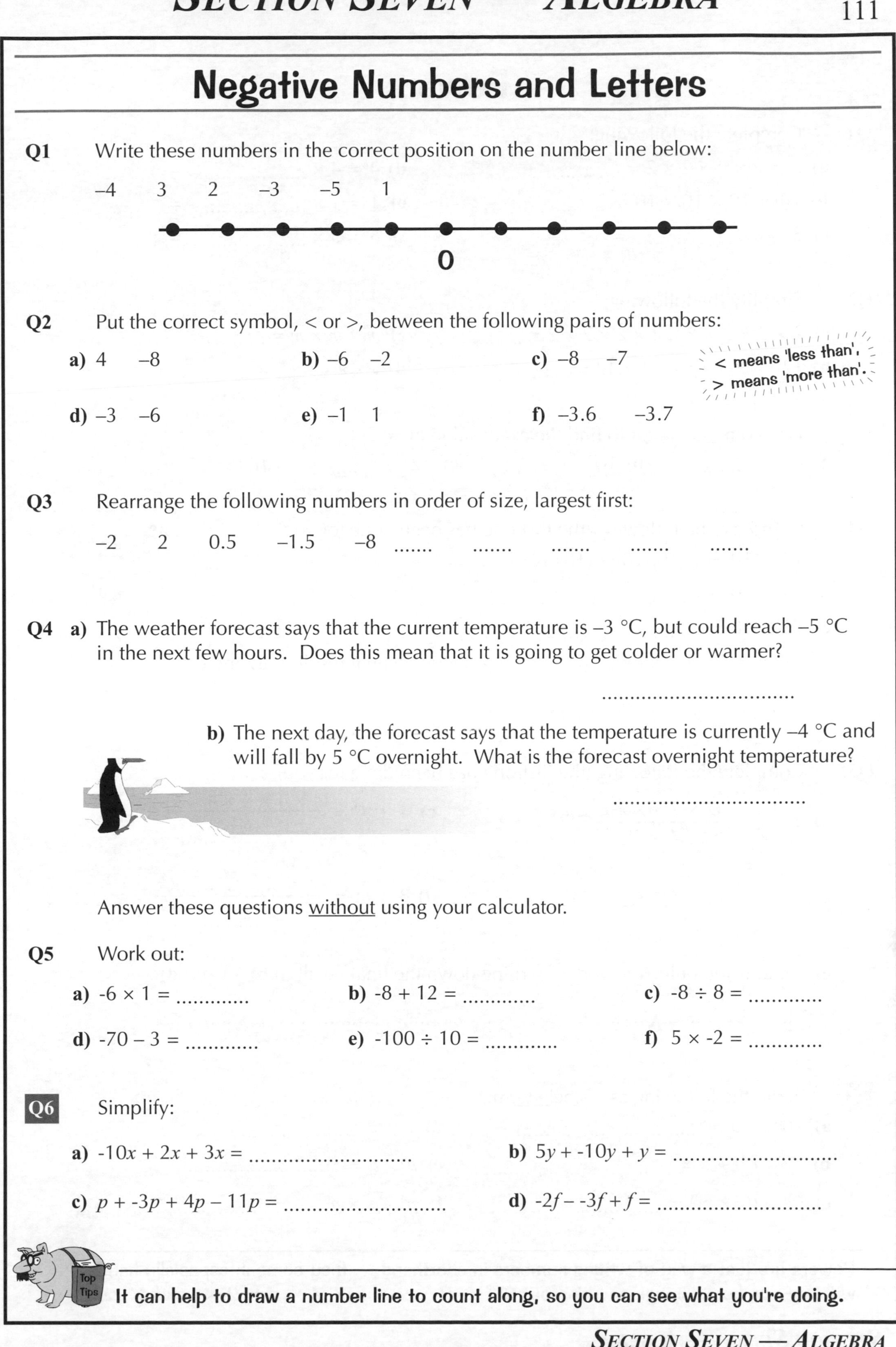

Negative Numbers and Letters

Q1 Write these numbers in the correct position on the number line below:

–4 3 2 –3 –5 1

0

Q2 Put the correct symbol, < or >, between the following pairs of numbers:

a) 4 –8 **b)** –6 –2 **c)** –8 –7

d) –3 –6 **e)** –1 1 **f)** –3.6 –3.7

< means 'less than',
> means 'more than'.

Q3 Rearrange the following numbers in order of size, largest first:

–2 2 0.5 –1.5 –8 ....

Q4 **a)** The weather forecast says that the current temperature is –3 °C, but could reach –5 °C in the next few hours. Does this mean that it is going to get colder or warmer?

....

b) The next day, the forecast says that the temperature is currently –4 °C and will fall by 5 °C overnight. What is the forecast overnight temperature?

....

Answer these questions <u>without</u> using your calculator.

Q5 Work out:

a) $-6 \times 1 =$ .... **b)** $-8 + 12 =$ .... **c)** $-8 \div 8 =$ ....

d) $-70 - 3 =$ .... **e)** $-100 \div 10 =$ .... **f)** $5 \times -2 =$ ....

Q6 Simplify:

a) $-10x + 2x + 3x =$ .... **b)** $5y + -10y + y =$ ....

c) $p + -3p + 4p - 11p =$ .... **d)** $-2f - -3f + f =$ ....

Top Tips

It can help to draw a number line to count along, so you can see what you're doing.

Powers

Q1 Complete the following:

a) $2^4 = 2 \times 2 \times 2 \times 2 =$ ....

b) $10^3 = 10 \times 10 \times 10 =$ ....

c) $3^5 = 3\times$ .... = ...

d) $4^6 = 4 \times$ .... = ...

e) $1^9 = 1 \times$ .... = ...

f) $5^6 = 5 \times$ .... = ...

Q2 Simplify the following:

a) $2 \times 2 \times 2 \times 2 \times 2 \times 2 \times 2 \times 2 =$ ....

b) $12 \times 12 \times 12 \times 12 \times 12 =$ ....

c) $m \times m \times m =$ ....

d) $y \times y \times y \times y =$ ....

Q3 Use your <u>calculator</u> to find the exact value of

a) 4^3 .... **b)** 10^4 .... **c)** 12^5 .... **d)** 13^3 ....

Q4 Complete the following (the first one has been done for you):

a) $10^2 \times 10^3 = (10\times10) \times (10\times10\times10) = 10^5$

b) $10^3 \times 10^4 =$ .... =

c) $10^4 \times 10^2 =$ .... =

d) What is the <u>quick method</u> for writing down the final result in **b)** and **c)**?

....

Q5 Complete the following (the first one has been done for you):

a) $2^4 \div 2^2 = \frac{(2 \times 2 \times 2 \times 2)}{(2 \times 2)} = 2^2$

b) $2^5 \div 2^2 = \frac{(2 \times 2 \times 2 \times 2 \times 2)}{(2 \times 2)} =$ ....

c) $4^5 \div 4^3 = \frac{(4 \times 4 \times 4 \times 4 \times 4)}{\text{......................}} =$ ....

d) $8^5 \div 8^2 = \frac{\text{......................}}{\text{......................}} =$ ....

e) What is the quick method for writing down the final result in **b)**, **c)** and **d)**?

....

Q6 Write the following as a <u>single term</u>:

a) $10^6 \div 10^4 =$ ....

b) $(8^2 \times 8^5) \div 8^3 =$ ....

c) $6^{10} \div (6^2 \times 6^3) =$ ....

d) $x^2 \times x^3 =$ ....

e) $a^5 \times a^4 =$ ....

f) $p^4 \times p^5 \times p^6 =$ ....

Powers are just a way of writing numbers in shorthand — they come in especially handy with big numbers. Imagine writing out 2^{138} — 2 × 2 ×... × 2 ×... × 2 × yawn × zzz...

Square Roots and Cube Roots

Square root just means "WHAT NUMBER TIMES ITSELF (i.e. 2×2) GIVES..."
The square roots of 64 are 8 and −8 because 8×8=64 and -8×-8=64.
Cube root means "WHAT NUMBER TIMES ITSELF TWICE (i.e. 2×2×2) GIVES ..."
The cube root of 27 is 3 because 3×3×3=27.
Square roots always have a + and − answer, cube roots only have 1 answer.

Tip

Q1 Use the $\sqrt{\ }$ button on your calculator to find the following positive square roots to the nearest whole number.

a) $\sqrt{60}$ = ...
b) $\sqrt{19}$ = ...
c) $\sqrt{34}$ = ...
d) $\sqrt{200}$ = ...
e) $\sqrt{520}$ = ...
f) $\sqrt{75}$ = ...
g) $\sqrt{750}$ = ...
h) $\sqrt{0.9}$ = ...
i) $\sqrt{170}$ = ...
j) $\sqrt{7220}$ = ...
k) $\sqrt{1\,000\,050}$ = ...
l) $\sqrt{27}$ = ...

Q2 Without using a calculator, write down both answers to each of the following:

a) $\sqrt{4}$ = ...
b) $\sqrt{16}$ = ...
c) $\sqrt{9}$ = ...
d) $\sqrt{49}$ = ...
e) $\sqrt{25}$ = ...
f) $\sqrt{100}$ = ...
g) $\sqrt{144}$ = ...
h) $\sqrt{64}$ = ...
i) $\sqrt{81}$ = ...

Q3 Use your calculator to find the following:

a) $\sqrt[3]{4096}$ = ...
b) $\sqrt[3]{1728}$ = ...
c) $\sqrt[3]{1331}$ = ...
d) $\sqrt[3]{1\,000\,000}$ = ...
e) $\sqrt[3]{1}$ = ...
f) $\sqrt[3]{0.125}$ = ...

Q4 Without using a calculator, find the value of the following:

a) $\sqrt[3]{64}$ = ...
b) $\sqrt[3]{27}$ = ...
c) $\sqrt[3]{1000}$ = ...
d) $\sqrt[3]{8}$ = ...

Q5 Nida is buying a small gift box online. She sees a cube box with volume of 125 cm^3. What is the length of each box edge?

....

Q6 A farmer is buying fencing to surround a square field of area 3600 m^2. What length of fencing does he need to buy?

....

Algebra

Algebra can be pretty scary at first. But don't panic — the secret is just to practise lots and lots of questions. Eventually you'll be able to do it without thinking, just like riding a bike. But a lot more fun, obviously...

Simplifying means collecting like terms together:	**Expanding** means removing brackets:
$8x^2 + 2x + 4x^2 - x + 4$ becomes $12x^2 + x + 4$ ($8x^2$: x^2 term; $2x$: x term; $4x^2$: x^2 term; $-x$: x term; 4: number term)	E.g. $4(x + y) = 4x + 4y$ $x(2 + x) = 2x + x^2$ $-(a + b) = -a - b$

Q1 By collecting like terms, simplify the following. The first one is done for you.

a) $6x + 3x - 5 = 9x - 5$

b) $2x + 3x - 5x =$ ....

c) $9f + 16f + 15 - 30 =$ ....

d) $14x + 12x - 1 + 2x =$ ....

e) $3x + 4y + 12x - 5y =$ ....

f) $11a + 6b + 24a + 18b =$ ....

g) $9f + 16g - 15f - 30g =$ ....

h) $14a + 12a^2 - 3 + 2a =$ ....

Q2 Simplify the following. The first one is done for you.

a) $3x^2 + 5x - 2 + x^2 - 3x = 4x^2 + 2x - 2$

b) $5x^2 + 3 + 3x - 4 =$ ....

c) $13 + 2x^2 - 4x + x^2 + 5 =$ ....

d) $7y - 4 + 6y^2 + 2y - 1 =$ ....

e) $2a + 4a^2 - 6a - 3a^2 + 4 =$ ....

f) $15 - 3x - 2x^2 - 8 - 2x - x^2 =$ ....

g) $x^2 + 2x + x^2 + 3x + x^2 + 4x =$ ....

h) $2y^2 + 10y - 7 + 3y^2 - 12y + 1 =$ ....

Q3 Expand the brackets and then simplify if possible. The first one is done for you.

Careful with the minus signs — they multiply both terms in the bracket.

a) $2(x + y) = 2x + 2y$

b) $4(x - 3) =$ ....

c) $8(x^2 + 2) =$ ....

d) $-2(x + 5) =$ ....

e) $-(y - 2) =$ ....

f) $x(y + 2) =$ ....

g) $x(x + y + z) =$ ....

h) $8(a + b) + 2(a + 2b) =$ ....

Algebra

FACTORISING is just putting the brackets back in.
And when you've just spent all that time getting rid of them...

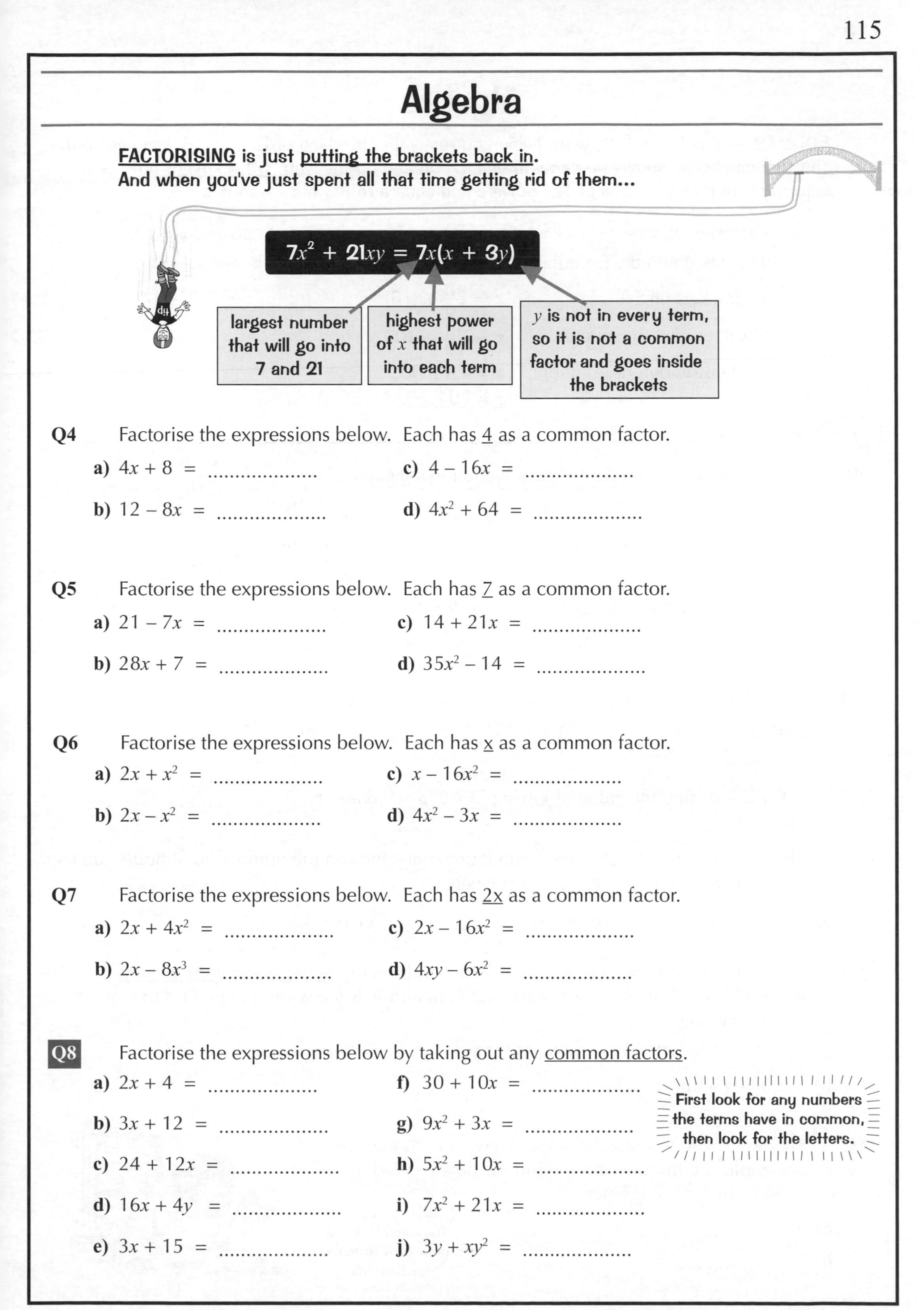

Q4 Factorise the expressions below. Each has 4 as a common factor.

a) $4x + 8$ = ... **c)** $4 - 16x$ = ...

b) $12 - 8x$ = ... **d)** $4x^2 + 64$ = ...

Q5 Factorise the expressions below. Each has 7 as a common factor.

a) $21 - 7x$ = ... **c)** $14 + 21x$ = ...

b) $28x + 7$ = ... **d)** $35x^2 - 14$ = ...

Q6 Factorise the expressions below. Each has x as a common factor.

a) $2x + x^2$ = ... **c)** $x - 16x^2$ = ...

b) $2x - x^2$ = ... **d)** $4x^2 - 3x$ = ...

Q7 Factorise the expressions below. Each has 2x as a common factor.

a) $2x + 4x^2$ = ... **c)** $2x - 16x^2$ = ...

b) $2x - 8x^3$ = ... **d)** $4xy - 6x^2$ = ...

Q8 Factorise the expressions below by taking out any common factors.

a) $2x + 4$ = ... **f)** $30 + 10x$ = ...

b) $3x + 12$ = ... **g)** $9x^2 + 3x$ = ...

c) $24 + 12x$ = ... **h)** $5x^2 + 10x$ = ...

d) $16x + 4y$ = ... **i)** $7x^2 + 21x$ = ...

e) $3x + 15$ = ... **j)** $3y + xy^2$ = ...

First look for any numbers the terms have in common, then look for the letters.

Algebra

BODMAS — this funny little word helps you remember in which order to work formulas out. The example below shows you how to use it. Oh and by the way "Other" might not seem important, but it means things like powers and square roots, etc — so it is.

Example: if $z = \frac{x}{10} + (y - 3)^2$, find the value of z when $x = 40$ and $y = 13$.

1) Write down the formula with the numbers stuck in $z = \frac{40}{10} + (13 - 3)^2$,

2) Brackets first: $z = \frac{40}{10} + (10)^2$

3) Other next, so square: $z = \frac{40}{10} + 100$

4) Division before Addition: $z = 4 + 100$

$\underline{z = 104}$

Q9 If $x = 3$ and $y = 6$ find the value of the following expressions.

a) $x + 2y$ .... **c)** $4(x + y)$ .... **e)** $2x^2$ ....

b) $2x \div y$ .... **d)** $(y - x)^2$ .... **f)** $2y^2$ ....

Q10 If $V = lwh$, find V when, **a)** $l = 7$, $w = 5$, $h = 2$ ....

b) $l = 12$, $w = 8$, $h = 5$ ....

Q11 Using the formula $z = (x - 10)^2$, find the value of z when:

a) $x = 20$ .... **b)** $x = 15$ .... **c)** $x = -1$ ....

Q12 If $V = u + at$, find the value of V when $u = 8$, $a = 9.8$ and $t = 2$. ....

Q13 The cost in pence (C) of hiring a sun lounger depends on the number (n) of hours you use it for, where $C = 100 + 30n$. Find C when,

a) $n = 2$ .... **b)** $n = 6$ .... **c)** $n = 3.5$ ....

Q14 The cost of framing a picture, C pence, depends on the dimensions of the picture. If $C = 10L + 5W$, where L is the length in cm and W is the width in cm, find the cost of framing:

a) a picture of length 40 cm and width 24 cm ....

b) a square picture of sides 30 cm. ....

Q15 The time taken to cook a chicken is given as 20 minutes per lb plus 20 minutes extra. Find the time needed to cook a chicken weighing:

a) 4 lb ....

b) 7.5 lb ....

You need to write your own formula for this one.

Number Patterns and Sequences

Q1 Draw the next two pictures in each pattern.
How many match sticks are used in each picture?

a)

b)

c)

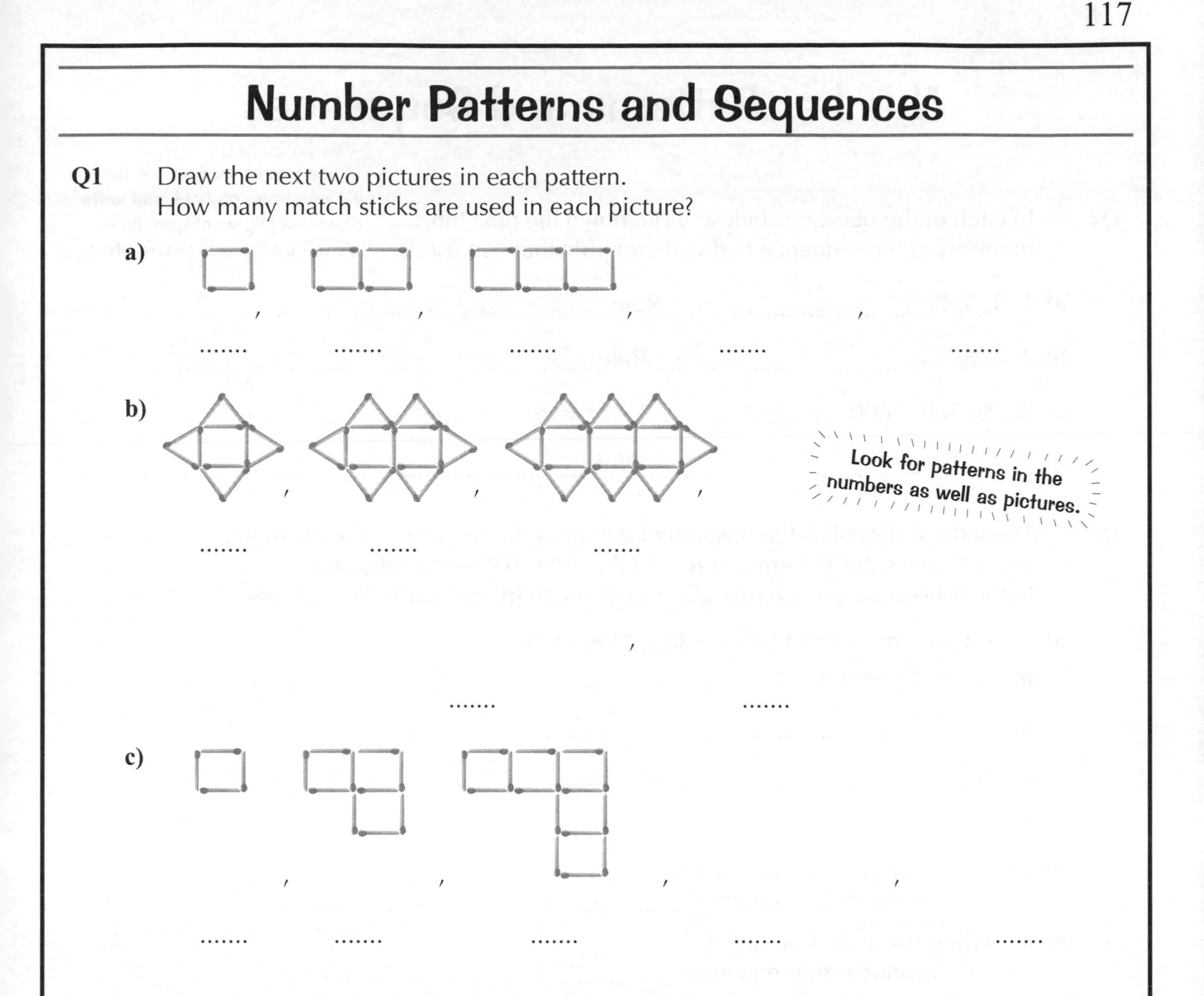

Q2 Look for the pattern and then fill in the next three lines. Some of the answers are too big to fit on a calculator display so you must spot the pattern.

a)

$7 \times 6 = 42$

$67 \times 66 = 4422$

$667 \times 666 = 444\ 222$

$6667 \times 6666 =$ ....

$66\ 667 \times 66\ 666 =$ ....

$666\ 667 \times 666\ 666 =$ ....

b)

$1 \times 81 = 81$

$21 \times 81 = 1701$

$321 \times 81 = 26\ 001$

$4321 \times 81 = 350\ 001$

$54\ 321 \times 81 =$ ....

$654\ 321 \times 81 =$ ....

$7\ 654\ 321 \times 81 =$ ....

Q3 The first five terms of a sequence are 3, 7, 11, 15, 19...
Is 34 a term in the sequence? Explain your answer.

....

Number Patterns and Sequences

Once you've worked out the next numbers, go back and write down exactly what you did — that will be the rule you're after.

Q4 In each of the questions below, write down the next three numbers in the sequence and write the rule that you used.

a) 1, 3, 5, 7, ... , ... , ... Rule ....

b) 2, 4, 8, 16, ... , ... , ... Rule ....

c) 3, 30, 300, 3000, ... , ... , ... Rule ....

d) 3, 7, 11, 15, ... , ... , ... Rule ....

Q5 The letter ***n*** describes the position of a term in the sequence. For example, if $n = 1$, that's the 1st term... if $n = 10$ that's the 10th term and so on.
In the following, use the rule given to generate (or make) the first 5 terms.

a) $3n + 1$ so if $n = 1$ the 1st term is **$(3 \times 1) + 1$** = **4**

$n = 2$ the 2nd term is .... = ...

$n = 3$ .... = ...

$n = 4$ .... = ...

$n = 5$ .... = ...

b) $5n - 2$, when $n = 1, 2, 3, 4$ and 5
produces the sequence .... , ... , ... , ... , ...

c) n^2, when $n = 1, 2, 3, 4$, and 5
produces the sequence .... , ... , ... , ... , ...

Q6 Write down an expression for the nth term of the following sequences:

Remember the formula — dn + (a – d).

a) 2, 4, 6, 8, ...

....

b) 1, 3, 5, 7, ...

....

c) 5, 8, 11, 14, ...

....

Q7 Mike has opened a bank account to save for a holiday. He opened the account with £20 and puts £15 into the account at the end of each week. So at the end of the first week the balance of the account is £35. What will the balance of the account be after:

a) 3 weeks? .... **b)** 5 weeks? ....

c) Write down an expression that will let Mike work out how much he'll have in his account after any number of weeks he chooses.

....

d) What will the balance of the account be after 18 weeks?

....

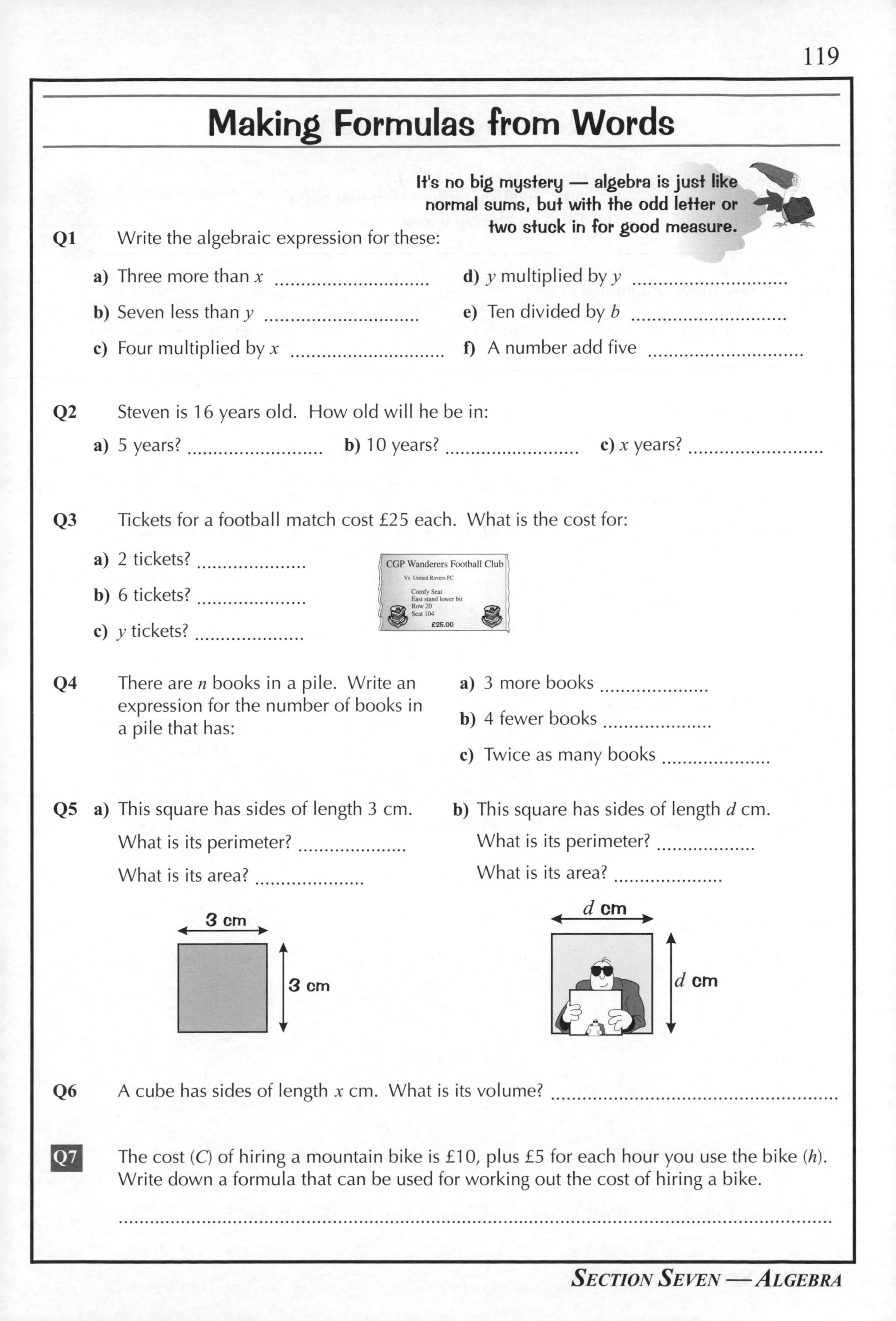

Making Formulas from Words

It's no big mystery — algebra is just like normal sums, but with the odd letter or two stuck in for good measure.

Q1 Write the algebraic expression for these:

a) Three more than x ....

b) Seven less than y ....

c) Four multiplied by x ....

d) y multiplied by y ....

e) Ten divided by b ....

f) A number add five ....

Q2 Steven is 16 years old. How old will he be in:

a) 5 years? ....
b) 10 years? ....
c) x years? ....

Q3 Tickets for a football match cost £25 each. What is the cost for:

a) 2 tickets? ....

b) 6 tickets? ....

c) y tickets? ....

Q4 There are n books in a pile. Write an expression for the number of books in a pile that has:

a) 3 more books ....

b) 4 fewer books ....

c) Twice as many books ....

Q5 **a)** This square has sides of length 3 cm.

What is its perimeter? ....

What is its area? ....

b) This square has sides of length d cm.

What is its perimeter? ....

What is its area? ....

Q6 A cube has sides of length x cm. What is its volume? ....

Q7 The cost (C) of hiring a mountain bike is £10, plus £5 for each hour you use the bike (h). Write down a formula that can be used for working out the cost of hiring a bike.

....

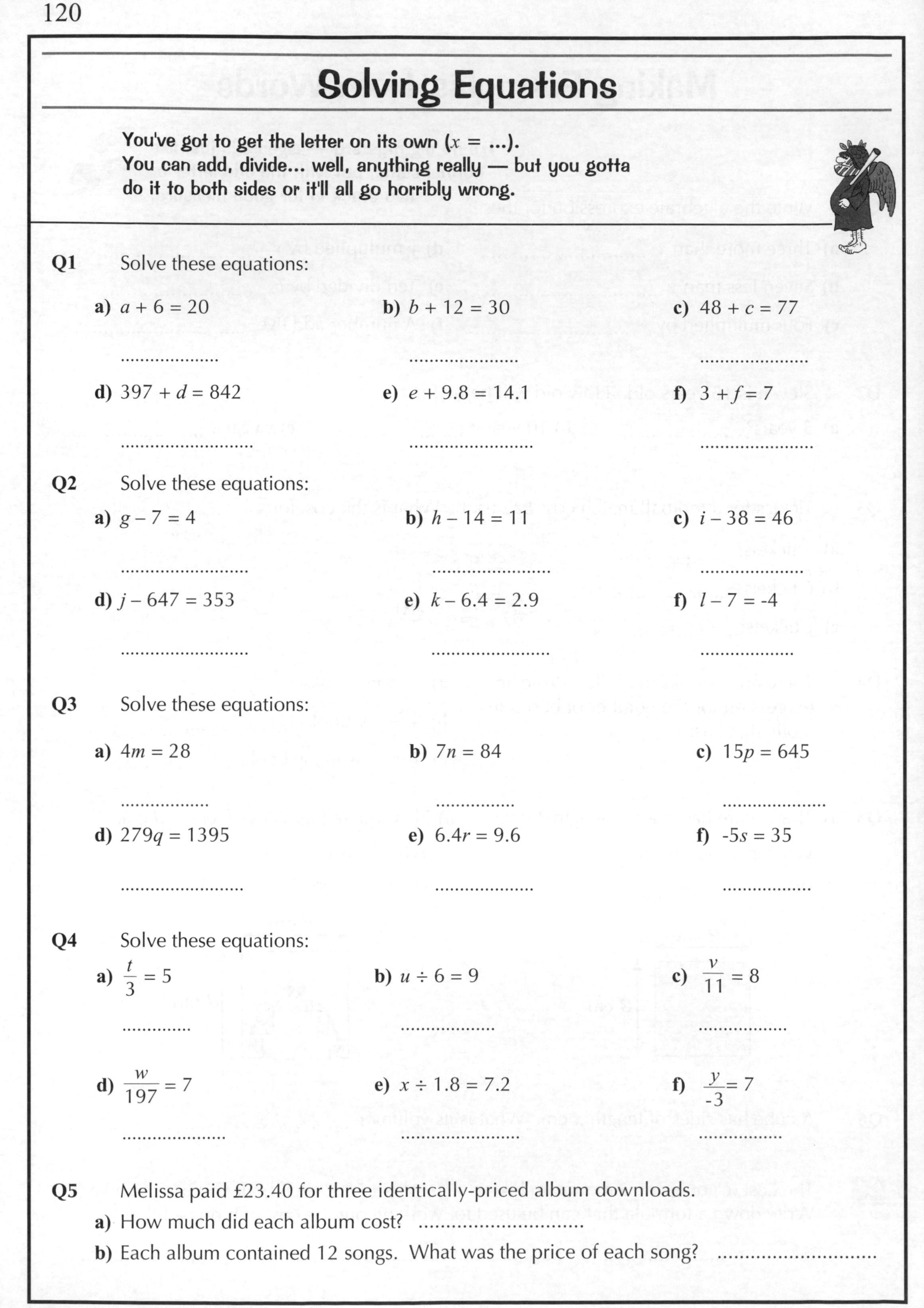

Solving Equations

You've got to get the letter on its own (x = ...). You can add, divide... well, anything really — but you gotta do it to both sides or it'll all go horribly wrong.

Q1 Solve these equations:

a) $a + 6 = 20$ ....

b) $b + 12 = 30$ ....

c) $48 + c = 77$ ....

d) $397 + d = 842$ ....

e) $e + 9.8 = 14.1$ ....

f) $3 + f = 7$ ....

Q2 Solve these equations:

a) $g - 7 = 4$ ....

b) $h - 14 = 11$ ....

c) $i - 38 = 46$ ....

d) $j - 647 = 353$ ....

e) $k - 6.4 = 2.9$ ....

f) $l - 7 = -4$ ....

Q3 Solve these equations:

a) $4m = 28$ ....

b) $7n = 84$ ....

c) $15p = 645$ ....

d) $279q = 1395$ ....

e) $6.4r = 9.6$ ....

f) $-5s = 35$ ....

Q4 Solve these equations:

a) $\frac{t}{3} = 5$ ....

b) $u \div 6 = 9$ ....

c) $\frac{v}{11} = 8$ ....

d) $\frac{w}{197} = 7$ ....

e) $x \div 1.8 = 7.2$ ....

f) $\frac{y}{-3} = 7$ ....

Q5 Melissa paid £23.40 for three identically-priced album downloads.

a) How much did each album cost? ....

b) Each album contained 12 songs. What was the price of each song? ....

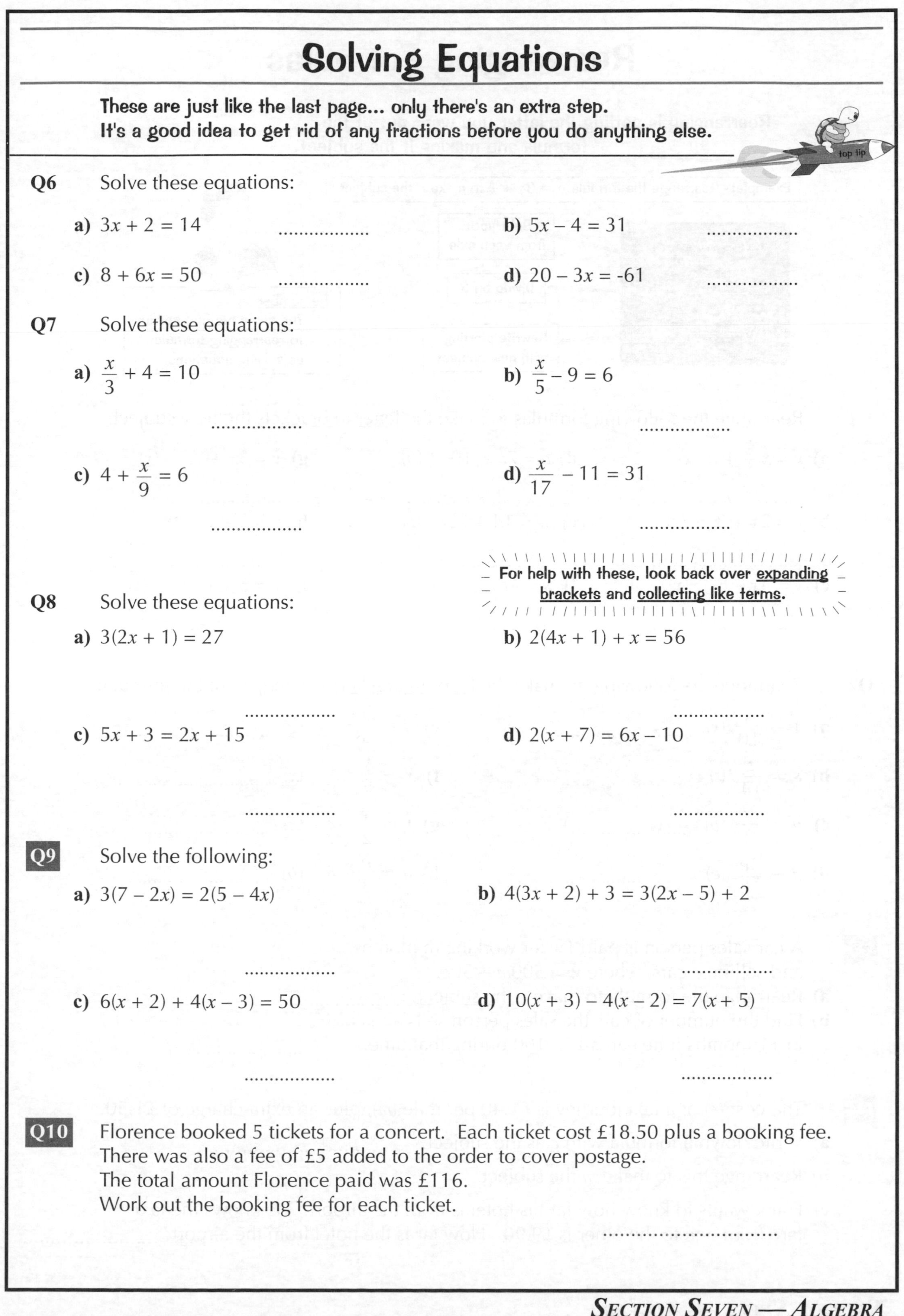

Solving Equations

These are just like the last page... only there's an extra step.
It's a good idea to get rid of any fractions before you do anything else.

Q6 Solve these equations:

a) $3x + 2 = 14$ ....

b) $5x - 4 = 31$ ....

c) $8 + 6x = 50$ ....

d) $20 - 3x = -61$ ....

Q7 Solve these equations:

a) $\frac{x}{3} + 4 = 10$ ....

b) $\frac{x}{5} - 9 = 6$ ....

c) $4 + \frac{x}{9} = 6$ ....

d) $\frac{x}{17} - 11 = 31$ ....

For help with these, look back over expanding brackets and collecting like terms.

Q8 Solve these equations:

a) $3(2x + 1) = 27$ ....

b) $2(4x + 1) + x = 56$ ....

c) $5x + 3 = 2x + 15$ ....

d) $2(x + 7) = 6x - 10$ ....

Q9 Solve the following:

a) $3(7 - 2x) = 2(5 - 4x)$ ....

b) $4(3x + 2) + 3 = 3(2x - 5) + 2$ ....

c) $6(x + 2) + 4(x - 3) = 50$ ....

d) $10(x + 3) - 4(x - 2) = 7(x + 5)$ ....

Q10 Florence booked 5 tickets for a concert. Each ticket cost £18.50 plus a booking fee.
There was also a fee of £5 added to the order to cover postage.
The total amount Florence paid was £116.
Work out the booking fee for each ticket.

....

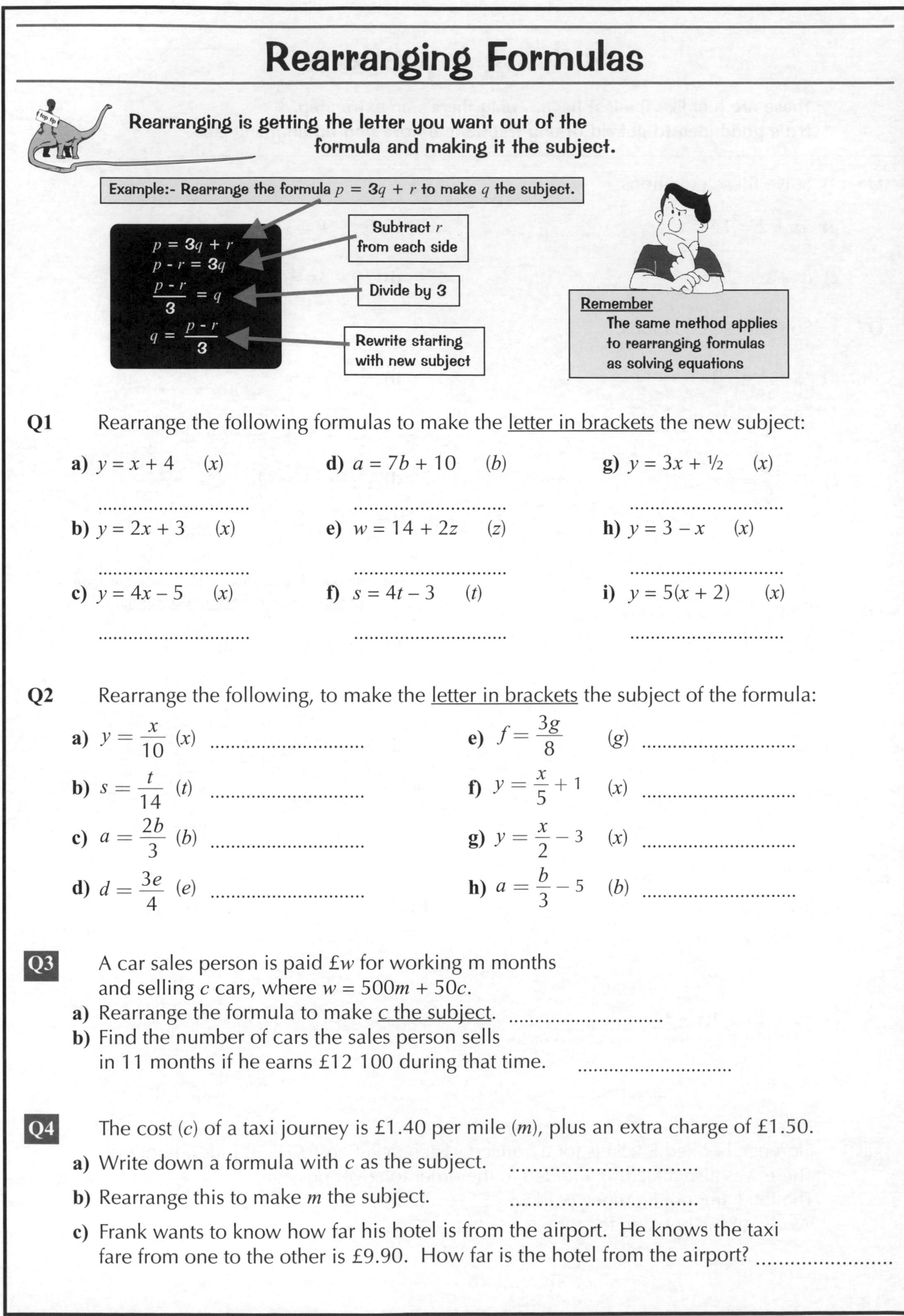

Rearranging Formulas

Rearranging is getting the letter you want out of the formula and making it the subject.

Example:- Rearrange the formula $p = 3q + r$ to make q the subject.

$$p = 3q + r$$
$$p - r = 3q$$
$$\frac{p - r}{3} = q$$
$$q = \frac{p - r}{3}$$

Subtract r from each side

Divide by 3

Rewrite starting with new subject

Remember
The same method applies to rearranging formulas as solving equations

Q1 Rearrange the following formulas to make the letter in brackets the new subject:

a) $y = x + 4$ (x) ....

b) $y = 2x + 3$ (x) ....

c) $y = 4x - 5$ (x) ....

d) $a = 7b + 10$ (b) ....

e) $w = 14 + 2z$ (z) ....

f) $s = 4t - 3$ (t) ....

g) $y = 3x + ½$ (x) ....

h) $y = 3 - x$ (x) ....

i) $y = 5(x + 2)$ (x) ....

Q2 Rearrange the following, to make the letter in brackets the subject of the formula:

a) $y = \frac{x}{10}$ (x) ....

b) $s = \frac{t}{14}$ (t) ....

c) $a = \frac{2b}{3}$ (b) ....

d) $d = \frac{3e}{4}$ (e) ....

e) $f = \frac{3g}{8}$ (g) ....

f) $y = \frac{x}{5} + 1$ (x) ....

g) $y = \frac{x}{2} - 3$ (x) ....

h) $a = \frac{b}{3} - 5$ (b) ....

Q3 A car sales person is paid £w for working m months and selling c cars, where $w = 500m + 50c$.

a) Rearrange the formula to make c the subject. ....

b) Find the number of cars the sales person sells in 11 months if he earns £12 100 during that time. ....

Q4 The cost (c) of a taxi journey is £1.40 per mile (m), plus an extra charge of £1.50.

a) Write down a formula with c as the subject. ....

b) Rearrange this to make m the subject. ....

c) Frank wants to know how far his hotel is from the airport. He knows the taxi fare from one to the other is £9.90. How far is the hotel from the airport? ....

Trial and Improvement

Q1 Use the trial and improvement method to solve the equation $x^3 = 50$.
Give your answer to one decimal place. Two trials have been done for you.

Try $x = 3$ $x^3 = 27$ (too small)
Try $x = 4$ $x^3 = 64$ (too big)

....

Q2 Use the trial and improvement method to solve these equations.
Give your answers to one decimal place.

a) $x^2 + x = 80$

....

b) $x^3 - x = 100$

....

Show all the numbers you've tried, not just your final answer... or you'll be chucking away easy marks.

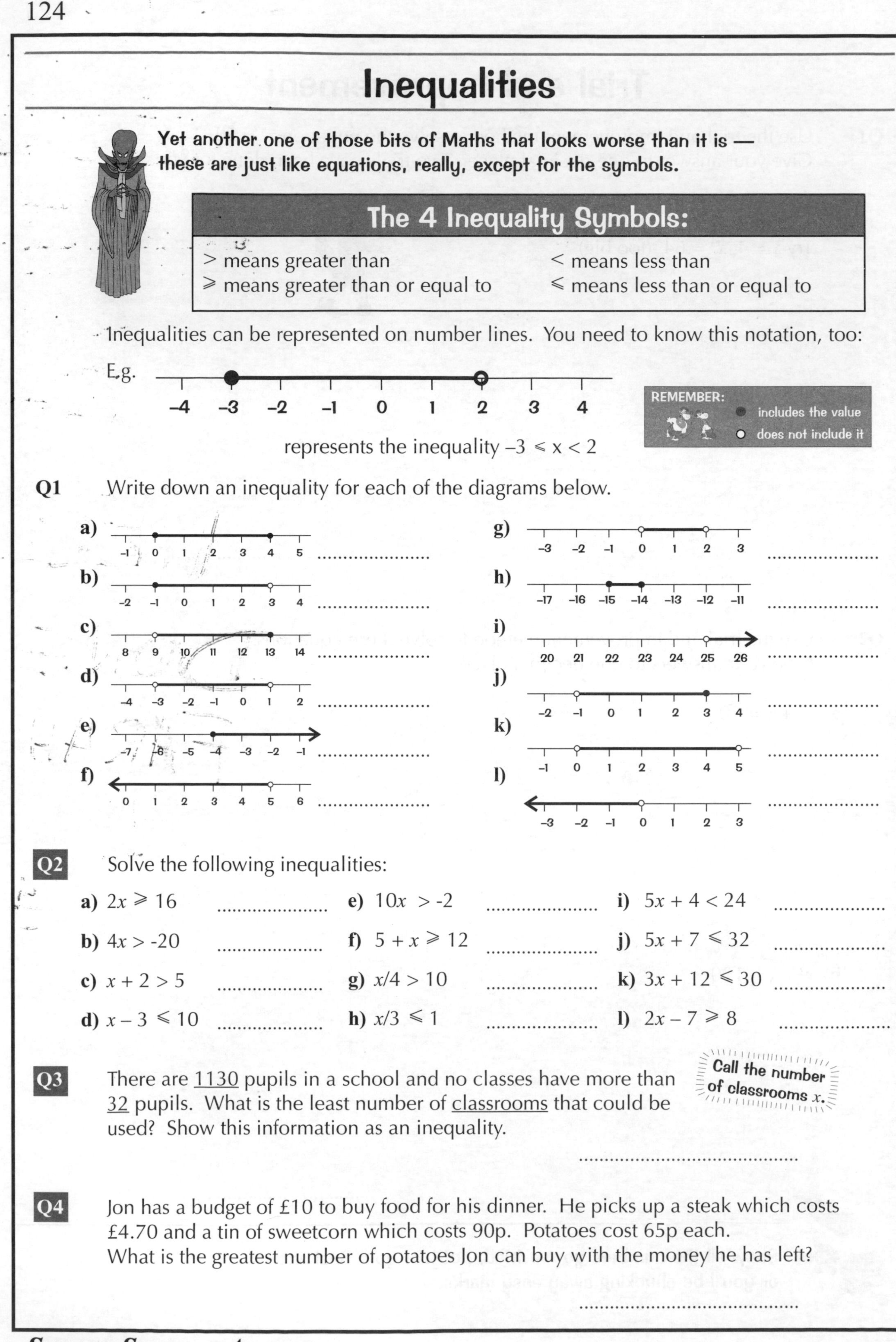

Inequalities

Yet another one of those bits of Maths that looks worse than it is — these are just like equations, really, except for the symbols.

The 4 Inequality Symbols:

$>$ means greater than	$<$ means less than
$\geqslant$ means greater than or equal to	$\leqslant$ means less than or equal to

Inequalities can be represented on number lines. You need to know this notation, too:

E.g.

represents the inequality $-3 \leqslant x < 2$

Q1 Write down an inequality for each of the diagrams below.

Q2 Solve the following inequalities:

a) $2x \geqslant 16$ ....

b) $4x > -20$ ....

c) $x + 2 > 5$ ....

d) $x - 3 \leqslant 10$ ....

e) $10x > -2$ ....

f) $5 + x \geqslant 12$ ....

g) $x/4 > 10$ ....

h) $x/3 \leqslant 1$ ....

i) $5x + 4 < 24$ ....

j) $5x + 7 \leqslant 32$ ....

k) $3x + 12 \leqslant 30$ ....

l) $2x - 7 \geqslant 8$ ....

Q3 There are 1130 pupils in a school and no classes have more than 32 pupils. What is the least number of classrooms that could be used? Show this information as an inequality.

Call the number of classrooms x.

....

Q4 Jon has a budget of £10 to buy food for his dinner. He picks up a steak which costs £4.70 and a tin of sweetcorn which costs 90p. Potatoes cost 65p each. What is the greatest number of potatoes Jon can buy with the money he has left?

....